Jossey-Bass Teacher

Jossey-Bass Teacher provides K–12 teachers with essential knowledge and tools to create a positive and lifelong impact on student learning. Trusted and experienced educational mentors offer practical classroom-tested and theory-based teaching resources for improving teaching practice in a broad range of grade levels and subject areas. From one educator to another, we want to be your first source to make every day your best day in teaching. *Jossey-Bass Teacher* resources serve two types of informational needs—essential knowledge and essential tools.

Essential knowledge resources provide the foundation, strategies, and methods from which teachers may design curriculum and instruction to challenge and excite their students. Connecting theory to practice, essential knowledge books rely on a solid research base and time-tested methods, offering the best ideas and guidance from many of the most experienced and well-respected experts in the field.

Essential tools save teachers time and effort by offering proven, ready-to-use materials for in-class use. Our publications include activities, assessments, exercises, instruments, games, ready reference, and more. They enhance an entire course of study, a weekly lesson, or a daily plan. These essential tools provide insightful, practical, and comprehensive materials on topics that matter most to K–12 teachers.

Hands-On Math Projects with Real-Life Applications

Second Edition

Judith A. Muschla
Gary Robert Muschla

JOSSEY-BASS
A Wiley Imprint
www.josseybass.com

Published by Jossey-Bass
A Wiley Imprint
989 Market Street, San Francisco, CA 94103-1741   www.josseybass.com

ISBN 13: 978-0-7879-8179-2
ISBN 10: 0-7879-8179-6

Jossey-Bass books and products are available through most bookstores. To contact Jossey-Bass directly call our Customer Care Department within the U.S. at 800-956-7739, outside the U.S. at 317-572-3986, or fax 317-572-4002.

Jossey-Bass also publishes its books in a variety of electronic formats. Some content that appears in print may not be available in electronic books.

Printed in the United States of America
SECOND EDITION
PB Printing   10 9

About the Authors

Judith A. Muschla received her B.A. in mathematics from Douglass College at Rutgers University and is certified to teach grades K–12. She has taught mathematics in South River, New Jersey, for over twenty-five years. She has taught math at various levels at South River High School, from basic skills through precalculus. She has also taught at South River Middle School where, in her capacity as a team leader, she helped revise the mathematics curriculum to reflect the Standards of the National Council of Teachers of Mathematics, coordinated interdisciplinary units, and conducted mathematics workshops for teachers and parents. She was a recipient of the 1990–1991 Governor's Teacher Recognition Program award in New Jersey and was named the 2002 South River Public School District Teacher of the Year. Along with teaching, she has been a member of the state Standards Review Panel for the Mathematics Core Curriculum Contents Standards in New Jersey.

Including this second edition of *Hands-On Math Projects with Real-Life Applications*, Judith and Gary Muschla have coauthored seven math books published by Jossey-Bass: *The Math Teacher's Book of Lists* (1995; 2nd edition, 2005), *Math Starters! 5- to 10-Minute Activities to Make Kids Think, Grades 6–12* (1999), *Geometry Teacher's Activities Kit* (2000), *Math Smart! Over 220 Ready-to-Use Activities to Motivate and Challenge Students, Grades 6–12* (2002), *Algebra Teacher's Activities Kit* (2003), and *Math Games: 180 Reproducible Activities to Motivate, Excite, and Challenge Students, Grades 6–12* (2004).

Gary Robert Muschla received his B.A. and M.A.T. from Trenton State College and taught in Spotswood, New Jersey, for more than twenty-five years. He spent many of his years in the classroom teaching mathematics at the elementary level. He has also taught reading and writing and is a successful author. He is a member of the Authors Guild.

He has written several resources for teachers, among them *The Writing Teacher's Book of Lists* (1991; 2nd edition, 2004), *Writing Workshop Survival Kit* (1993; 2nd edition, 2005), *English Teacher's Great Books Activities Kit* (1994), *Reading Workshop Survival Kit* (1997), *Ready-to-Use Reading Proficiency Lessons and Activities, 4th Grade Level* (2002), *Ready-to-Use Reading Proficiency Lessons and Activities, 8th-Grade Level* (2002), and *Ready-to-Use Reading Proficiency Lessons and Activities, 10th-Grade Level* (2003), all published by Jossey-Bass. He currently writes and serves as a consultant in education.

For Erin

Contents

Part One: Implementing Projects in the Math Class

Part Two: The Projects

Section One: Math and Science

Section Two: Math and Social Studies

Section Three: Math and Language

Section Four: Math and Art and Music

Section Five: Math and Sports and Recreation

Section Six: Math and Life Skills

Acknowledgments

We thank Michael J. Pfister, assistant superintendent of South River Public Schools, Kevin W. Kidney, principal of South River High School, Paul J. Coleman, assistant principal of South River High School, Geraldine Misiewicz, math supervisor, and our colleagues for their support of our writing.

Thanks also to Steve D. Thompson, our editor, for his support of our efforts to complete this second edition.

Special thanks to Caroline Fitzgerald and Geri Priest, who read the original manuscript and offered many helpful suggestions, to Jamie Egan for helping us better understand the mathematical aspects of music, and Colleen Duffey Shoup as well as Dover Clip-Art for the illustrations.

We greatly appreciate the help and encouragement of Susan Kolwicz, whose advice on the first edition enabled us to take our rough ideas and fashion them into a practical resource for teachers.

We are indebted to Sonia Helton, professor of education at the University of South Florida, for her insightful recommendations for the projects in this book.

We want to thank our daughter, Erin, who read the first draft of this new edition from the perspective of a young math teacher and caught several oversights and omissions.

And finally, we thank our students. In the end, they are why all of us are in this business.

About This Book

Appropriate for grades 6 through 12, *Hands-On Math Projects with Real-Life Applications, Second Edition* consists of two parts: Part One focuses on implementation and management, and Part Two contains sixty projects for your students. The projects support the Standards of the National Council of Teachers of Mathematics (NCTM), as well as meet the mandates of the No Child Left Behind Act that call for project-based learning, problem-solving strategies in mathematics, and the integration of technology in the classroom.

The new edition of this book retains the valued features of the first edition, while updating its relevance and extending its scope. All of the projects have been revised to reflect current trends, numerical data have been redone, and information on the use of technology has been greatly expanded. For many projects, students will use the resources of the World Wide Web to obtain information that they then use to solve the problem the project presents. Some projects have been replaced with new projects that hold greater applications for today, including Project 25, "Rating Math Web Sites," Project 55, "Maintaining a Math Class Web Site," and Project 56, "Selecting a Sound System Using the Internet."

The new edition, like the original one, is designed for easy implementation. Each project stands alone and may be used with students of various grade levels and abilities, providing teachers with great flexibility for instruction.

To prepare students for the demands they will face in the workplace, math teachers must provide a classroom environment where students are challenged to solve real-life problems, where they may collaborate and share ideas, where they use calculators and computers, where they express their thoughts orally and in writing, and where they recognize that mathematics is not an isolated subject but is connected to other disciplines. The projects in this book will help you to achieve these goals.

Alignment to National Council of Teachers of Mathematics Standards

The following table indicates the NCTM Standards addressed by the projects of this book. Checks indicate the specific standards with which each project aligns. For some projects, skills will vary depending on the material students create for the project.

Project Number	Number and Operations	Algebra	Geometry	Measurement	Data Analysis and Probability	Problem Solving	Reasoning and Proof	Communication	Connections	Representation
1					√	√	√	√	√	√
2					√	√	√	√	√	√
3	√			√	√			√	√	√
4	√	√		√	√	√		√	√	√
5		√	√					√		√
6	√		√	√		√	√	√	√	√
7	√	√	√	√	√	√	√	√	√	√
8	√	√		√	√	√		√	√	
9				√	√			√	√	√
10			√	√		√	√	√		√
11						√	√	√	√	
12	√				√	√	√	√	√	√
13		√	√							√
14	√		√	√						√
15	√	√						√	√	√
16								√	√	√
17	√				√	√	√	√	√	√
18				√	√				√	√
19	√			√	√			√	√	√
20						√		√	√	√
21							√	√	√	√
22						√	√	√	√	√

Project Number	Number and Operations	Algebra	Geometry	Measurement	Data Analysis and Probability	Problem Solving	Reasoning and Proof	Communication	Connections	Representation
23						√	√	√	√	√
24						√	√	√	√	√
25					√	√	√	√	√	√
26								√	√	
27						√		√	√	
28						√	√	√	√	√
29								√	√	√
30								√	√	√
31						√	√	√	√	√
32						√	√	√	√	√
33								√		√
34						√	√	√	√	√
35		√	√	√		√			√	√
36								√	√	√
37						√		√	√	√
38	√					√	√	√	√	√
39		√	√	√					√	√
40				√				√		√
41			√					√	√	√
42			√	√		√			√	√
43	√				√	√	√	√	√	√
44	√					√	√	√	√	
45	√				√	√	√	√	√	√
46	√				√	√	√	√	√	√
47	√	√		√		√	√	√	√	√
48	√			√		√	√	√	√	√
49								√	√	
50	√			√		√	√	√	√	
51	√	√		√	√	√	√	√	√	
52	√				√	√	√	√	√	√
53	√		√	√		√		√	√	√
54	√					√	√	√	√	√
55					√	√	√	√	√	√
56	√					√	√	√	√	
57	√	√				√	√	√	√	
58						√	√	√	√	√
59	√					√		√	√	√
60						√	√	√	√	√

How to Use This Resource

Hands-On Math Projects with Real-Life Applications, Second Edition is divided into two parts. Part One, "Implementing Projects in the Math Class," contains three chapters devoted to classroom management. Part Two, "The Projects," offers sixty math projects designed to enhance your math program.

Before assigning any of the projects, we recommend that you read through Part One in its entirety because the information it provides will help you to implement project activities in your class. After reading Part One, you may select those projects in Part Two that best support your program and satisfy the needs of your students.

In Part One, Chapter One provides an overview of how to incorporate math projects in your class, Chapter Two details a variety of specific classroom management techniques and suggestions, and Chapter Three offers several methods for evaluating the work of your students. Each of the chapters includes several lists that summarize information for various topics, making it easy for you to find the information you need. For example, you will find that your role expands when projects become part of your program. "The Teacher's Role During Math Projects" outlines the many tasks you may assume when your students are engaged in project work.

Each chapter of Part One also includes several reproducibles for students that can be helpful in establishing the routines necessary for successful project work. For example, it is possible that many students may not have had much experience in working cooperatively to solve a complex problem. Distributing copies of "Rules for Working in Math Teams" highlights the behaviors that characterize effective teams. Knowing what is expected of them helps students to behave appropriately and achieve the goals you set out for them.

Part Two contains sixty projects divided into six major sections:

Section One, Math and Science

Section Two, Math and Social Studies

Section Three, Math and Language

Section Four, Math and Art and Music

Section Five, Math and Sports and Recreation

Section Six, Math and Life Skills

Although the breakdown is useful for planning interdisciplinary units or finding a project that ties in to another subject, each project stands alone. Each may be used to introduce, enhance, or conclude a unit or topic, or be used as a challenge, enrichment, or extra credit. Some projects may be used as ongoing activities—for example, Project 31, "Keeping a Math Journal," or Project 32, "Math Portfolios." Project 24, "The Mathematics Publishing Company," shows students how to create and produce a mathematics magazine, which you may decide to publish regularly throughout the year.

Each project follows the same format. Information for the teacher is presented first: background, goals of the project, math skills that are covered, special materials and equipment that are needed, and development. This material is followed by the Student Guide, which provides strategies and suggestions on how students may solve the problem the project presents. Data sheets and worksheets provide students with additional information or a specialized work space. The student guides, data sheets, and worksheets, which are numbered according to each project, are reproducible for your convenience.

We suggest that you use this book as a resource, selecting projects you need to enhance your curriculum. The sixty projects offer a variety of real-life situations that will help your students to realize the relevance math has in their lives, while at the same time applying specific mathematics skills.

We trust that you will find this book to be a helpful resource as you encourage and support your students in their efforts to learn math. Our best wishes to you.

Part One

Implementing Projects in the Math Class

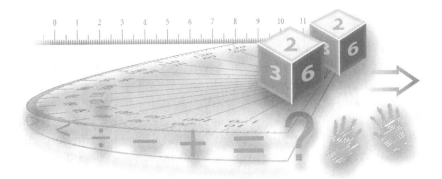

Chapter One

Overview of projects in the math class

In a well-run math class, computation, problem solving, and critical thinking are all taught. Instead of learning skills in isolation, students learn math in context where they can see how it is applied in real situations. In this way they come to recognize the importance of math in their own lives. The connection between math and the real world is a strong one. This is especially true of math classes in which projects are a significant part of the curriculum.

Filled with activity and enthusiasm, a successful project-oriented math class is a center of individual learning, collaboration, cooperation, and sharing. Students work alone, together, and with the teacher. Along with learning fundamental math skills, students learn to think logically, analyze data, make decisions, and solve multifaceted problems that arise out of real-life situations. Students thus use the skills they are learning in meaningful ways.

Your Role

Your role changes when your students work on math projects. Along with your traditional responsibilities of introducing concepts, demonstrating skills with example problems, and grading the work of your students, you will become a facilitator and promoter. The horizons of your teaching will expand. More of your time will be spent working directly with individuals and groups. As students work on solving problems, you will circulate around the room, offering advice and suggestions, asking questions that lead to insights or direction, and giving encouragement and praise. Sometimes you may simply monitor a group's efforts or model appropriate behavior. Occasionally you may need to pull a group back on task. (See "The Teacher's Role During Math Projects.")

There are many ways you can incorporate projects into your curriculum. While following your text, you can easily provide regular project activities. You may build time for projects into your schedule, for example, a day or two each week, or do units on projects a few times a year. Some teachers introduce a multistep project and then give students time to work on it at the end of class over the next few days. No matter how you provide the time, you should be consistent. Students not only need sufficient time for working on projects, they need to know when they will be working on them. This information enables students to come to class prepared and ready to work.

Supporting the Standards of the National Council of Teachers of Mathematics

The projects in this book support the Standards of the National Council of Teachers of Mathematics (NCTM), with specific emphasis on the following:

- The Problem-Solving Standard
- The Communication Standard
- The Connections Standard
- The Representation Standard

Problem solving is an essential part of learning mathematics. At its most basic, it requires students to find a solution to a problem without initially knowing what methods or procedures to use. As they seek a solution to the problem, students must draw on their own knowledge, experience, and skills. They may be required to assume various tasks, including conducting research, analyzing and organizing data, and drawing conclusions. In their efforts to discover answers, they reinforce previously learned skills, acquire new skills, and gain a greater understanding of mathematics. By presenting students with a variety of practical, engaging problems to solve, the projects contained in this book support the Problem-Solving Standard and foster the learning of valuable problem-solving skills.

The projects also support the Communication Standard. Because effective communication depends on clear thought and expression, communication encourages students to think critically, formulate their ideas, and express those

The Teacher's Role During Math Projects

Since discovery is an important part of any project, you must encourage your students to assume much of the responsibility for their learning and progress. Your role changes. Along with your traditional duties, you will be spending some of your class time doing many of the following:

- Presenting multistep, critical thinking projects based on real-life situations

- Organizing and monitoring groups so that members work effectively together

- Modeling appropriate behavior and problem-solving skills

- Demonstrating to students what it is to be an enthusiastic problem solver by showing them how you are willing to tackle projects that at first may seem impossible

- Brainstorming with groups

- Guiding students in their research efforts

- Showing students that process is crucial to finding solutions

- Offering suggestions to solve problems

- Offering encouragement and applauding efforts

- Explaining that mistakes are merely stepping-stones to finding solutions and to learning

- Answering questions

- Helping students sort through their thoughts as they consider problem-solving strategies

- Showing students that various strategies may be used to solve the same problem

- Providing sufficient time for working on projects

- Monitoring student behavior and ensuring that classroom procedures are followed

- Keeping students on task

- Evaluating and assessing student progress

- Providing time for sharing results

ideas with mathematical precision. Communication gives students the opportunity to state their ideas, listen to the ideas of others, and compare them to their own, furthering their understanding of math. An important part of every project of this book is sharing results through formal and informal presentations.

Along with the Standards for Problem Solving and Communication, working on math projects supports the Connections Standard. Although students are often taught mathematical skills in isolation or in packets of information, mathematics is a broad, complex subject in which ideas are interconnected, extending throughout the field of math and to other disciplines. As they work on projects, students will find relationships among ideas that will broaden their understanding of problems and solutions, thereby gaining an appreciation of the scope of mathematics and how math is interwoven through all parts of society.

The projects of this book also support the Representation Standard. Mathematical ideas are represented with notations, symbols, and figures. Typical examples of representations include numbers, expressions, diagrams, and graphs. There are many more, of course. As students work on projects, they will express mathematical ideas as representations, which then become tools to explore, model, and develop mathematical concepts. An understanding of mathematical representations will serve students well in their continued study of math.

Perhaps one of the greatest benefits of using math projects in the classroom is how students must draw on numerous skills as they work toward solutions. When students work on math projects, they expand their view of math to real-life situations and develop skills that stretch well beyond the traditional curriculum. Such results support the Standards of the NCTM and enrich the mathematical experiences of students.

Strategies for Problem Solving

The math projects your students will be doing will require computation, analysis, problem-solving and critical thinking skills, and decision making. Since the type and nature of problems vary, there can be no set plan or step-by-step process they can use all the time. You should familiarize your students with various strategies that they can draw on as needed. Emphasize that strategies are methods or procedures that can be used alone or with other strategies. If a student should ask what strategy is best for solving a particular problem, a good answer is, "The one that works best for you." You will likely find that different students will use different strategies to solve the same problem.

While some students may be quite adept at problem solving, many will need guidance, and you may wish to distribute copies of "Problem-Solving Strategies." It is a guide that can help students get started in solving problems and keep them moving along.

There is also much you can do in regular lessons to help students acquire sound problem-solving skills that will be useful to them throughout their lives. See "Helping Students Develop Problem-Solving Skills" for a list of suggestions.

Problem-Solving Strategies

There are many ways to solve multistep problems. If you believe that there can be only one or two, you limit your options and reduce your chances of finding a solution. Following are some suggestions and strategies.

Before you begin seeking the solution:

- Make sure that you understand the problem. This may require rereading it several times.
- Be sure you understand the question and what answers you are seeking.
- Look for "hidden" questions.
- Find the important information that the problem provides, and eliminate information that is not essential. (Sometimes problems contain facts that you do not need.)
- Supply any missing information. You may need to research and analyze data.
- Make sure that you understand any special facts, data, or units of measurement.

As you seek a solution, consider all of these strategies:

- Look for patterns, relationships, connections, sequences, or causes and effects.
- Use guess and check (also called trial and error). Choose a place to start, try a solution, and see if it works. If it does not, try another.
- Organize your facts and information in a list. Sometimes this exercise can show relationships that you might otherwise overlook.
- Construct a table or chart. This is another way of identifying relationships.
- Think logically. Look for sequence and order.
- Rely on common sense. Some answers simply are not possible. Do not waste time pursuing them.
- Sketch or draw a model to help you visualize the problem.
- Simplify the problem by breaking it down into manageable steps. Solve a sub-problem that leads to the solution of a bigger problem.
- Look at the problem from different angles.
- Estimate. Rounding numbers can make it easier to find a solution. Using whole numbers rather than fractions may help you to see operations more clearly.
- Act the problem out.
- Keep notes of your attempted solutions. This will reduce the chances that you will repeat steps that do not move you forward.
- Periodically review your notes and attempts at solutions. By rechecking what you have done, you might see something you overlooked.
- Do not give up. The persistent problem solver finds solutions.

When you believe you have found the answer:

- Double-check your work.
- Be certain that you used all necessary information.
- Recheck your calculations.
- Be sure that your answer is logical.

Helping Students Develop Problem-Solving Skills

You can help your students learn critical thinking and problem-solving skills by doing the following:

- Present students with real-life problems to which they can relate.

- Offer problems that have multiple solutions and can be solved through several strategies.

- Encourage your students to try various strategies in solving problems.

- Organize students into cooperative teams.

- Encourage students to brainstorm for ideas that might lead to solutions.

- Give problems that have missing information *and* too much information. Such problems require students to supply and eliminate data.

- Give problems that tie in to other subjects.

- Encourage students to keep logs or notes of their efforts at solving difficult problems.

- Encourage students not to give up; persistence is a major factor in successful problem solving.

- Require students to write explanations of how they solved problems.

- Remind students to always check answers for logic and accuracy.

- Encourage discussion and the sharing of solutions.

A vital part of any project is the sharing of solutions and results at the end of the activity. When results are shared, students have the opportunity to hear other viewpoints, learn about other methods used to solve problems, and realize that others may have experienced some of the same stumbling blocks they did. Not only does this help reduce an individual's feelings that he or she is the only one having trouble, it also helps build a sense of class community and problem-solving camaraderie.

Sharing may be oral through presentations using technology such as interactive whiteboards or Microsoft PowerPoint, or written in the form of logs or reports. Thus, speaking and writing become essential components of your math class.

Perhaps the biggest factor that holds many students back from becoming good problem solvers is a lack of confidence. Many students doubt that they can solve complex problems and give up with little effort. Explain to your students that problem-solving skills come with practice. Just like anything else—learning to play a musical instrument, excelling at gymnastics, playing computer games—the more they work at solving problems, the better they will become. Distribute copies of "What It Takes to Become a Top Problem Solver" to highlight some of the characteristics that successful problem solvers share. The list can serve as a guide, detailing traits and attitudes that your students should strive to acquire throughout the year.

Creating Your Own Projects

While this book provides projects that require various steps and strategies, you may eventually wish to create projects of your own, designed specifically for your students. Material for math projects is all around you. As you develop projects, keep in mind the following points, which will help ensure that your projects are stimulating and exciting to your students:

- Base your projects on real-life situations that are meaningful to your students.
- Design projects that capture the interest of your students.
- Make sure that your students possess the mathematical skills to solve the problems they will encounter in your projects.
- Develop projects that require analysis, critical thinking, and decision making.
- Create projects that require students to formulate a plan to find a solution.

For suggestions where you can find material from which to create projects for your students, see "Sources for Developing Math Projects."

What It Takes to Become a Top Problem Solver

Top problem solvers share many of the same traits. You can become a successful problem solver. All it takes is practice. The more problems you solve, the more skilled you will become. Try to make the following traits part of your personality.

Good problem solvers are:

- Confident that they can solve just about any problem.

- Persistent in solving problems.

- Willing to try different strategies to solve problems.

- Able to find important information and eliminate unimportant facts.

- Able to recognize patterns, relationships, and connections.

- Able to look at a problem from various viewpoints.

- Open to new ideas.

- Willing to make notes to keep track of their attempts at solutions.

- Able to draw on other experiences in the solving of problems.

- Able to use logic and common sense.

Sources for Developing Math Projects

Good material for creating your own math projects is all around you. The following sources are particularly useful.

- Your math text likely contains sections such as "Challenges" that offer interesting facts or situations that you can easily turn into fine projects. Some texts have sections of data banks that provide information that can serve as the basis for projects.

- The Internet offers vast information on countless topics. Information can be easily found by conducting a simple search by topic.

- National and local newspapers contain an assortment of valuable information. Charts and tables can be especially helpful.

- Regional and national magazines are good sources of information for projects.

- Almanacs and other reference books can provide unusual and interesting data on countless topics.

- Major events at school can be your springboard for creating projects. Use field day, homecoming, the Valentine's Day dance, or the prom to capture the interest of your students.

- Books of math puzzles and games frequently offer a wealth of ideas for projects.

- Consult with your colleagues and develop projects that include two or more subject areas. Science and social studies in particular share many topics with mathematics.

Conclusion

Without question, math projects offer many benefits to students. Perhaps most important, when students work on authentic problems, they see how the math skills they are learning may be applied to the real world. Math projects open the door to bringing other subjects and disciplines into the math class, and students quickly recognize that math is interwoven through many parts of their lives. Math projects also give students the opportunity to work together cooperatively, share their experiences, and celebrate the solving of problems that might be overwhelming for one person to manage. Furthermore, when students collaborate on a project, students of all abilities have the chance to contribute to the solution. Everyone has a part to play; everyone has a role to fill; everyone can be a contributor to and a sharer in success.

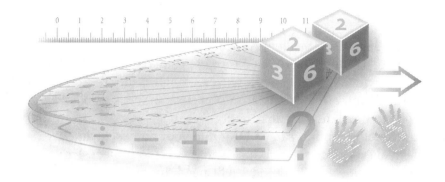

Chapter Two

Managing projects in your math class

A successful math class in which projects are an important part of the course of study is the result of effective planning and management. Along with teaching required material, you must provide meaningful projects that have real-life applications. These are not easy charges. As students work on projects, they will be engaged in various tasks: they will need to consider different strategies, gather and analyze information, confer with each other, manipulate models, perform calculations, and test possible solutions. All this requires an environment that promotes vigorous inquiry, encourages students to assume the responsibility for their learning, and supports both individual and group activities.

Structuring Your Class

Math projects provide students a chance to use various skills in solving authentic problems. Since projects often reach beyond the math class, they offer an excellent way to broaden the scope of your curriculum and introduce exciting new activities to your teaching. There are many ways you can incorporate projects into your classes. Perhaps the easiest is to select projects that support the unit you are teaching. For example, if you are studying a unit in geometry, Project 6, "Designing a Flower Bed," in which students work with rectangles, squares, circles, and scale, will be useful. If you are teaching a unit on data analysis, Project 12, "An Election Poll," will supplement your instruction.

Projects used to enhance a unit can be built into your daily schedule. We suggest that you take a class period to introduce and begin the project. Explain the project; distribute any materials students might need; organize teams; and give students twenty minutes or so to plan, brainstorm potential strategies, and get started.

After this introductory period, resume your regular lessons, reserving about twenty minutes at the end of each period for students to continue working on the project. Providing students with time at the end of class eliminates the need for them to arrange to meet outside class (although sometimes students become so involved with a project that they meet on their own). This also allows you to assign the ordinary amount of homework and continue moving forward with the unit. Since not all teams will finish at the same time, groups that finish early may work on extensions of the project, write in math journals, or simply do homework. When every team is done, you should schedule a period for sharing results.

Another way to incorporate projects into your classes is to periodically set aside time for them. After you complete a unit, take three or four days to work on a project. Some teachers prefer this method because it gives students a break from the routines of the class but does not interfere with the general curriculum.

Perhaps you will decide to build projects into your schedule. You might reserve every Tuesday and Friday for working on projects. This plan has the advantage of establishing a regular schedule that ensures time for project work. Since the projects in this book stand alone, each can be used at any time during the year and still provide the benefits of using a variety of skills in meaningful contexts.

Some projects, especially those for individuals, may be completed at home. You might give students a choice of these projects as either assignments or bonus activities. Although the projects are completed on the students' time, you should provide class time for sharing. Responses to their results are important to students.

When selecting projects for your students, be sure to consider their needs and abilities. Never assign projects that require skills your students have not yet mastered. Students will find such projects to be a frustrating struggle, and any new skills they acquire will be offset by the negative emotions they come to feel for math.

Without question, math projects offer students several benefits. They permit students to use many skills in solving various problems, give students a chance to take ownership of their mathematics learning, and help students to see the relevance between math and real life.

Creating and Maintaining a Positive Environment

Problem solving thrives in an environment in which people work on problems that have valid applications to life, feel free to risk making mistakes, and are encouraged to share their ideas. The best problem solving occurs in classes where students enjoy the freedom to pursue learning in their own way. The tone you set in your classroom, your expectations, and the procedures you maintain are the foundation for such an environment.

Since students usually rise, or fall, to a teacher's expectations, always discuss your goals for the class with your students at the beginning of the year. Share with them how you intend to conduct the class, how the class will be organized, and what will be covered. Note what you expect from them.

For students to work efficiently on math projects, they need a classroom that is logistically comfortable for problem solving. Tables are ideal; however, if you do not have tables, you can push desks together. Either way, you should provide enough room between teams so that they can function as single entities without distractions from other groups. Along with enough space between teams, there should be enough work area for students to discuss possible strategies with each other, confer about data, manipulate and examine models, and work on calculations.

Support problem solving in whatever ways you can. Bulletin boards, corridor display cases, media center exhibits, and math fairs are just some ways to draw attention to your program. Always look for ways to highlight lists of problem-solving strategies, interesting articles about math, and the work of your students.

While there is much you can do to promote success in your classes, students too must strive to make the class beneficial. This is particularly true during group activities. Your students must be willing to accept more responsibility than is demanded by the traditional class. During project activities, they must remain focused on the tasks. Group work is not a time to talk about who might be named homecoming king or queen. Distributing copies of "The Responsibilities of the Math Student" to your students is an excellent way to share basic expectations for student behavior.

Unquestionably, student learning flourishes in a conducive environment. One of your most important tasks as a teacher is to create a classroom filled with enthusiasm, the spirit of inquiry, and the desire to learn. The best classes are founded on the spirit of cooperation and energetic intellectual pursuit, in which students believe that everyone can learn (and enjoy!) math. For a summary of characteristics of math classes that have a positive atmosphere, see "The Right Environment."

The Responsibilities of the Math Student

A successful math class results when people work together to learn math. Accepting the following responsibilities is the first step to making this class worthwhile:

- Each day report to class on time and ready to work.

- Remember to bring your text, notebooks, pencils, calculators, and other materials to class.

- Pay attention in class, and ask questions when you do not understand something.

- *Everyone can learn math.* Work hard, and finish your class work and homework.

- Work cooperatively with other students in groups. Share your ideas, and be willing to listen to the ideas of others.

- Try various strategies in solving problems.

- Remember that solving complicated, multistep problems takes time. Be persistent.

- Follow the classroom rules and procedures.

- Behave properly.

- Recognize the importance of math in your life.

The Right Environment

The following characteristics are found in math classes described as having a positive atmosphere:

- The goals of the class are high enough so that students have to work hard, but not so high that they feel frustrated with math and its applications.
- The classroom is built on openness, fresh ideas, and sharing.
- Teacher and students believe that everyone, regardless of gender and ethnicity, can learn math.
- Students' work is prominently displayed.
- The classroom is designed to support inquiry and problem solving.
- The classroom is bright and cheerful.
- The classroom adheres to orderly procedures. Students maintain appropriate behavior and follow the classroom rules.
- Goals and objectives are clear to students.
- Classroom rules are fair and consistent.
- The grading system is reasonable and equitable.
- The teacher interacts with students and is a guide, nurturer, cheerleader, and provider of information.
- Teachers model problem-solving behavior and share with students their own enthusiasm for finding solutions.
- Math is connected to real-life problems and situations.
- Cooperation is encouraged.
- Enough time is provided for problem solving.
- Students are encouraged to consider and explain their reasoning during problem solving.
- Students are encouraged to use various strategies in solving problems. They come to recognize that the same problem may have many solutions.
- Sharing is encouraged, especially how students found solutions to problems.
- Calculators, computers, and other technologies are used regularly in class.
- Math is related to other subjects as much as possible.
- Manipulative materials are used whenever possible to show students relationships.
- Students learn the value of mathematics in their lives.
- Students and teachers become partners in learning mathematics.

Teaching Suggestions

While every teacher has his or her individual techniques and methods, we have found that the following plan is helpful in presenting projects and problem solving. It can be broken down into three parts: introduction, work time, and wrapping up.

Introduction

Begin a math project by presenting the situation and problems that are to be solved. Offer examples, review any concepts or specific skills that students will need to solve the problems they will confront, and relate the project to real-life scenarios as much as possible. Encourage students to ask questions. Having a student paraphrase the project and what needs to be solved can be helpful in clarifying what everyone is to do.

Once students understand the project, distribute copies of student guide sheets and discuss the information presented there. Data sheets and additional materials, if any, should also be distributed. Having everything they need to begin helps students to see the full scope of the project.

Work Time

As students work in teams, your task is to circulate around the room, offering help, encouragement, or simply observing. This is also a time to monitor and model student behavior.

Pay close attention that a team does not stray off the topic. If you see this happening, you might point out their mistake or nudge them in the right direction. However, avoid giving answers to any problems. If students feel that you will provide answers, they will be less inclined to do the hard thinking that will result in finding answers themselves. To encourage students to find their own answers, some teachers insist that they (the teachers) may be asked a question only after the question has been presented to the team and no one else is able to answer it.

As you observe students, you may find that a team has trouble starting. Sometimes this is caused by students' not being able to focus the problem. Have students restate the problem and break it down into parts, concentrating their efforts to identify the most important facts. Teams may also have trouble finding strategies that will lead to solutions. In this case, suggest that teams brainstorm various strategies, and examine each one to see if it leads to a possible solution.

As you move around the room, be aware of the interactions of the members of each team. You will likely see that some groups work well together with everyone sharing ideas, others are dominated by one or two members, and some are just unmotivated. When a team is working well, leave it alone. Even offering a comment might disrupt its momentum. Remember that a project is a time for students to discover their own solutions. If a group is not working well, you should sit in on it and model appropriate behavior. Make sure that everyone is participating, and encourage team members to help each other. If necessary, for a time, assume the role of team leader to get things going, then gradually fade into the background as students begin to assume ownership of the project. With some teams, you may need to remind students of the proper procedures and behavior often, especially during the first few weeks of class.

Wrapping Up

Sharing is essential to the successful culmination of a math project. Discussing methods and results helps students to realize that some problems have multiple solutions that may be discovered through various strategies. This is an important lesson of authentic problem solving. In the real world, many problems have several solutions and can be solved in many ways. For more information on sharing, see "The Importance of Sharing" presented later in this chapter.

Individual and Team Conferences

As students work on projects, you will monitor the progress of the teams. In many cases, they will have questions, or you will need to discuss procedures, rules, or behavior. You will undoubtedly be conducting conferences with individuals or the entire team.

A conference does not have to be long; in fact, it may last only a minute or two. In most cases, it will be conducted at the students' work area. The purpose of any conference is to help students better understand the project they are working on, as well as help them to improve their understanding of mathematics. Often you may find that students need you only to answer a simple question. In such instances, provide guidance and let them get back to work. If an individual or team seems stuck, use this as your starting point for the conference.

Focus any conference on a particular problem or skill. If you try to do too much, you will confuse students or provide them with too much information. Either way, you will end their efforts to solve the problem. During the conference, be sure to keep your tone positive, and offer specifics. You may need to point a team in the right direction to find more information, offer encouragement to the team that is about to give up, or assure a team that their efforts are worthwhile.

When you give praise, it should be genuine, because students can tell when it is not. Always avoid negative or sarcastic remarks, for these will only discourage students. The conference should be a time of help and support.

The Value of Cooperative Problem Solving

In many jobs, people work in teams, and the experience your students gain now as they work together on math projects will serve them not only in your class but in the future as well. Teamwork fosters inquiry and discussion, and students often learn more when working together than they do trying to solve a complicated problem alone. Cooperative learning also provides students with the opportunity to acquire valuable social skills.

When students work in teams, they are more likely to take an active role. It is easier for them to get involved because the team provides support to individuals. Seeing other team members struggling with the same problems helps students feel less intimidated about offering their thoughts, and many students who would not risk sharing ideas with the whole class usually will share with their team. Furthermore, when they offer suggestions toward the solution of a problem, they receive immediate feedback. This sharing frequently results in a free-wheeling give-and-take of mathematics that is as stimulating as it is useful.

As a team works on a math project, it becomes involved in various activities. Team members need to discuss and assign tasks, reflect on how to approach the problem, test strategies, gather and analyze data, reach solutions, and determine how to justify and share results. Teamwork helps build student confidence, promotes critical thinking, and results in a sense of ownership of the problem.

Organizing Teams

Random groups tend to make the best math teams, although you should reserve the right to make adjustments. Groups of four to six generally work well for complex projects. With fewer than four, it is sometimes hard to generate enough ideas, especially if one of the students is absent or is shy or quiet.

An easy way to make random pairings is to simply count down your roster in sets of five, assigning the numbers 1 to 5 to students. Then all of the number 1s in the class would be on team 1, all of the 2s on team 2, and so on. Before announcing the teams to the students, review them and make sure that you have a mix of students of high and low abilities, as well as a mix of gender and ethnicity. It is also often a good idea to avoid having best friends or students who do not get along on the same teams. Make any final changes before informing students about the groups.

You should also change teams periodically. Rearranging groups allows students to interact with various personalities and see different viewpoints. In the real world, individuals are often required to work with people of varying outlooks and abilities. When making new teams, you can easily switch two members from each of the existing teams. Switching only one member keeps too much of the original team together.

After you have arranged your teams, explain to your students the purpose of working together. Suggest that a team may work most efficiently when tasks are divided. You may also suggest that students assume various roles that will help define responsibilities. For example, one student might serve as team leader. Her purpose is to keep the group on task and guide it toward the solution of the problem. Another might be the recorder, whose responsibility includes writing down the team's ideas, strategies, and conclusions. The list "How to Set up Project Teams" provides suggestions for organizing groups.

Unless your students have worked in teams before, they will probably need training in the procedures of effective teamwork. You should focus much of your attention on team interaction during the first project. You will likely need

How to Set up Project Teams

Working together in teams offers students an excellent way to learn math. The following guidelines can help you to organize your math teams:

- For complex projects, groups of four to six work best for middle and high school students.

- Organize your teams randomly; however, be sure to mix abilities, genders, ethnicities, and personalities.

- To build team spirit, suggest that teams select a name and design a team logo.

- Rearrange teams periodically. This gives students the chance to interact with others and experience new working relationships.

- Always explain the purpose of group work and expected behaviors. Some students may have little experience working in teams.

- Since teams often benefit from a division of labor, consider having students assume specific roles—for example:

  Leader, who guides the team toward its goal and makes sure everyone stays on task

  Recorder, who keeps notes of the team's ideas, strategies, and solutions

  Time monitor, who keeps track of time and helps the leader keep the team moving

  Checker, who reviews the work of the team

  Materials monitor, who assumes responsibility for any materials the team uses

  Presenter, who shares the team's findings with others.

- In small groups, students may assume more than one role.

Rules for Working in Math Teams

The success of a math team depends on the ability of its members to work together. Keeping the following suggestions in mind can help you and your team work more efficiently.

Each member of the team:

- Is responsible for his or her own behavior.

- Should work with other team members.

- Should help other members.

- Should share his or her ideas.

- Should carefully consider his or her ideas before speaking.

- Should give the floor to others after speaking.

- Should listen carefully and politely when others are speaking.

- Should ask questions when he or she does not understand something.

- Should strive to keep the discussion on the project and keep comments constructive.

- Should keep his or her emotions in check. When disagreements arise, they should be discussed calmly.

- Should carry out his or her role in the group the best he or she can.

to model behavior and remind students of procedures often, especially at the beginning of the year. Sit in on the various groups and show them how to act and behave. Acquiring the skills necessary for effective group work may take students a few weeks, and distributing copies of "Rules for Working in Math Teams" can be helpful in discussing expected behavior.

The Importance of Sharing

Sharing is crucial to projects and problem solving. Becoming aware of other strategies and solutions can broaden students' understanding of math. Sharing may take the form of an oral report, a presentation incorporating the use of technology such as an interactive whiteboard or Microsoft PowerPoint, or a written log or summary.

At the end of a project, you should provide time for teams to share their methods and findings. For oral sharing, the student designated as presenter shares the team's results with the rest of the class. Encourage students to discuss successful strategies, as well as earlier strategies that they attempted but which did not work. It is possible that other teams tried the same strategies but got different results or experienced different problems. The more that math is discussed, the more opportunities students have to gain new insights.

After the presentation, encourage questions from the class. Do not permit questions during sharing, because the presenter might become distracted and may not cover all of the essential points. During questioning, other members of the team may help the presenter, but only one student should speak at a time. This is also the time for members of other teams to offer comments or observations. Emphasize that any discussion should be positive, and do not allow sarcastic or negative statements.

Sometimes presenters may need help to cover all the issues. Guide students to report the strategies they used, their methods, procedures, and solutions. If a student becomes blocked, a helpful question from you can get him started again. Consider asking questions like the following:

"How did you divide tasks in your group?"

"What was your initial plan?"

"What strategies did you consider?"

"What problems or obstacles did you run into?"

"What kinds of data did you need to gather?"

"What sources did you use for finding information?"

"How do you know your solution is valid?"

"Are there other possible solutions? If yes, what made you select one over the others?"

At the end of the session, summarize the project and the results that the teams obtained. Highlight any unusual strategies or problems encountered, and be sure to discuss how the mathematics applies to real life.

Writing in Math Class

The benefits of writing in math class are well documented. Writing provides students with a method through which they can examine and share their thoughts about mathematics in a formalized manner. Through writing, students can connect concepts they have already learned with new ideas, summarize their understanding of math, and communicate their thoughts to others. Few will argue that only when we truly understand something can we explain it and put it clearly into words.

Many types of writing can be done in math class. Some of the most common are:

- Writing about specific problems
- Summaries or reports
- Biographies of famous mathematicians
- Word problems for other students
- Publication of a mathematics magazine (see Project 24)
- Keeping a math journal (see Project 31)
- Maintaining a math portfolio (see Project 32)

Whenever your students write in math class, encourage them to share their thoughts and information about mathematical concepts, methods, and applications. Avoid allowing students to write about math in ways that show little thought, purpose, or insight. Students should select meaningful topics on which they can share information and their ideas.

It has been found that people write according to a process that has been aptly named the writing process. It is likely that the English teachers in your school are familiar with it. You might consult with your students' English teacher, who may be willing to support your efforts for having students write in math class. To help your students understand the writing process, distribute copies of "The Writing Process and Math," and discuss the stages of the process with your students.

Using Technology with Math Projects

Technology such as calculators and computers is essential for teaching, learning, and doing mathematics. Such devices are especially helpful for work on multifaceted math projects because they enable students to collect, organize, and analyze data; view dynamic images of mathematical models; and perform computations with accuracy and efficiency. We live in a technological world, and the technologically related skills your students learn in school will serve them throughout their lives.

Hands-on math projects with real-life applications

The Writing Process and Math

When you are writing articles in math, it will be helpful to follow the stages of the writing process. You have probably learned about this in your English classes. Writing can be broken down into various stages, or steps. Authors go through these steps when they write, moving back and forth through the various stages as necessary. Understanding this process can help you with your writing. Following are the stages of the writing process.

Stage 1: Prewriting
- Thinking of a purpose
- Generating ideas
- Brainstorming
- Researching and gathering facts
- Analyzing ideas
- Organizing ideas
- Focusing ideas

Stage 2: Drafting
- Writing
- Rearranging information and ideas as needed
- Expanding ideas

Stage 3: Revising
- Rewriting
- Rethinking, rearranging, deleting, adding
- Clarifying ideas
- Checking ideas and mathematical facts
- Conducting more research
- Redrafting

Stage 4: Editing
- Proofreading
- Making any final corrections, including those relating to math

Stage 5: Publishing or sharing
- Sharing your written work with others
- Producing copies of your work
- Displaying your work

Calculators

Calculators are an essential component of any math class. Freeing students from the slow work of manual computation, they provide a means of proficient computing and allow more time for investigation, reasoning, decision making, and problem solving.

Calculators are particularly helpful for students who might have difficulty with computation. These students often become so worried about the basic operations that they do not enjoy the many benefits of project work such as collaboration, gathering and analyzing data, and making decisions. Without question, calculators are critical to your students in their work on math projects, because they enable students to focus on problem solving rather than computation.

Computers

Like calculators, computers are vital to modern math classes and assume a significant role in project work. Computers can be used to gather, analyze, and organize data; provide visual images of mathematical concepts and ideas; and link students to the Internet and mathematical Web sites throughout the world. They can support student studies and investigations of all areas of math and provide a foundation on which problem solving flourishes. Moreover, they can be used to print results and are essential in presenting the conclusions of a project using applications such as PowerPoint and interactive whiteboards.

PowerPoint Presentations

Microsoft's PowerPoint and similar applications enable users to design and give presentations, ranging from the basic to the sophisticated. Such presentations can be a fine culminating activity for your students' work on math projects. PowerPoint is not difficult for students to use, and mastery of it will serve students not only in your class but in other classes as well.

A typical PowerPoint presentation requires a computer that is running PowerPoint software, a display screen, and a projector. At its simplest, the information appearing on the computer's monitor is projected onto the screen. The presentation itself may contain text and images, including photographs, illustrations, tables, spreadsheets, and graphs. The images, which may be created with other applications or may be downloaded from the Internet or other media, are easily inserted into PowerPoint. Written material is simple to enter, arrange, and edit, commands being similar to word processing software. Presentations can be set up and run according to a timed sequence, or they can be manually controlled by use of a mouse. Once they understand the basics of the program, both high school and middle school students will find a PowerPoint presentation to be relatively easy to create and present.

PowerPoint presentations provide an organized visual as well as auditory learning experience and are particularly suited to sharing the results of a math project. Rather than simply reporting their findings to the class, students can present their results in a meaningful manner using PowerPoint.

Hands-on math projects with real-life applications

Interactive Whiteboards

Interactive whiteboards have been in use since the early 1990s. Models today, also known as electronic whiteboards and digital whiteboards, encompass a whiteboard (which is an electronic display), a computer, and a digital projector. A basic interactive whiteboard system enables its user to create, arrange, and manipulate data on the electronic display. Typical features include:

- Writing over information or objects displayed on the board
- Moving objects
- Rotating objects
- Resizing or recoloring objects
- Inserting backgrounds or images
- Saving files
- Sharing files
- Changing to the next screen when used with PowerPoint

As the original whiteboard technology has advanced, the software and interplay of the equipment have become seamless and relatively easy to use. Interactive whiteboards can help make the presentations of the results of math projects appealing and interesting.

Technology Training

While many, if not most, of your students will likely possess general computer skills, you should assume that most of them will not have had much experience with applications such as Microsoft's PowerPoint or interactive whiteboards. Yet this technology can support and enhance the work of your students on math projects. Incorporating it in your program offers significant benefits. Not only will your students learn the skills necessary to use the equipment, but they will be able to share ideas in an effective and interesting manner.

If your school has a computer specialist who works with students, meet with her and discuss the needs of your students. Perhaps she can incorporate the use of PowerPoint and interactive whiteboards in her instructional program. If your school does not have a technology or computer specialist who works with students, you must provide your students with the training they need to incorporate the use of such technologies with their project work. This is not as difficult as it may sound.

Although you need not be a technology expert, you should be familiar enough with the equipment so that you can demonstrate its basic use to your students. If necessary, ask the tech person at your school to show you how to use the equipment and software. Do not overlook the fact that most technology comes with tutorials that can show you the basics, and many vendors offer free technical support.

If you have the equipment in your classroom, set aside a period or two to provide students with basic instruction. If you do not have the equipment in your classroom, perhaps your school has a computer room or a section of the library in which computers containing PowerPoint or interactive whiteboards may be used. Reserve the equipment, and demonstrate its use to your students.

Start with a general overview, explaining the purpose and benefit of the equipment. For example, PowerPoint is an excellent tool for sharing the findings of a group, particularly if the presentation is to be done in a slide show format. If the presentation requires that images and information need to be moved or highlighted, the use of an interactive whiteboard may be more practical.

Guide students through the basics, explaining and demonstrating the most important commands. You might then have students come up in groups for more specific instruction. Base the amount of detail of your instruction on the abilities of your students. For most students, advanced commands are unnecessary. Approach the use of technology as a learning experience; there is no need to try to do more than students can comfortably manage.

Many students today are computer savvy and are able to learn the operations of new equipment quite readily. Consider organizing students into groups in a manner that places one or two technically inclined students with some whose technical skills may be weaker. The students with technical skills will take the lead with technology, reducing any anxiety other members of the group might have in regard to working with the equipment. Monitor the groups closely to ensure that the less technically minded do in fact gain experience working with the equipment and applications.

When it comes time to create a presentation of the results of project work, encourage students to work in school as well as at home (provided they have compatible software, which in many cases they will). They should save their material on a disk or CD, which they will then be able to use on the equipment in class. *A word of caution:* whenever students are bringing disks or CDs from home, using disks or CDS with computers in other classrooms, or even e-mailing files from home to school, the chance increases that your equipment could become infected with a computer virus. Always maintain updated virus protection software.

Depending on the abilities of your students, you may find it practical to work with individual groups as they set up presentations in the class. You can provide the guidance students will need when using equipment for the first time. Note, however, that not every project needs to culminate in a presentation using PowerPoint or an interactive whiteboard. To manage the workload, consider limiting major presentations to two or three groups per marking period. This permits you to spend more time with the groups. Keep track of the groups so that by the end of the year, everyone has a chance to use technology in their projects.

Math Projects and the Internet

The Internet contains thousands of Web sites devoted to mathematics, ranging from the history of math and basic operations to calculus and imaginary numbers and beyond. While some projects in this book have a specific technological component, for example, Project 25, "Rating Math Web Sites," Project 55, "Maintaining a Class Math Web Site," and Project 56, "Selecting a Sound System Using the Internet," many others require extensive use of the Internet for investigation.

Searching the Internet for Math

Because of the vast number of Web sites available, it is necessary for students to understand how to conduct practical searches of the Internet for math topics. Being able to search the Internet for information is an essential skill that will be useful for students far beyond math class.

Unless a person knows the Web address of a particular site (called its URL) that is likely to provide the information he seeks, he will have to conduct a search to find relevant information using a search engine, a sophisticated program designed to find information on the World Wide Web. Some of the most widely used search engines are:

Google	www.google.com
Webcrawler	www.webcrawler.com
All the Web	www.alltheweb.com
Alta Vista	www.altavista.com
Ask Jeeves	www.ask.com
Excite	www.excite.com
Lycos	www.lycos.com
Yahoo!	www.yahoo.com
Yahooligans	www.yahooligans.com

You can find more search engines at www.allsearchengines.com.

One of the best search engines for students, especially those in middle school, is Yahooligans. Yahooligans, a special part of Yahoo! is designed for young people. Instruct your students to go to the site and click on School Bells, then click on Math. A list of some of the best math sites for students (as determined by Yahoo editors) will be displayed.

If you would like to find more math Web sites, do a general search with one of the other search engines, for example, using the search term "math for students," and numerous sites will be offered.

Using Search Engines Effectively

The best search engine is only as effective as the precision of the search terms used. General search terms yield general results. They can result in so much data that the person searching for information finds it difficult, if not impossible, to locate the information she needs. Offer your students an example. Suppose a student is searching for information about quadrilaterals, a topic in geometry. If she searches using the term *geometry*, she will be presented with close to 2 million Web sites (we were when we conducted this search), all of which will provide information on geometry but not necessarily quadrilaterals. If she focuses her topic on quadrilaterals, she will be offered about twenty thousand Web sites, most of which are likely to offer information on her topic. Twenty thousand sites is still an unwieldy number, but 2 million is *a lot* more, and students will grasp the point of the example. Moreover, since most search engines list the most popular Web sites first, the chances of finding useful information with the first few hits are increased.

The following suggestions can make researching on the Internet more effective:

- Enter precise search terms and words.
- Use single words or precise topics when possible.
- Be specific. "Pascal's triangle" is more specific than "triangles," for example.
- Enter multiple spellings of words when applicable.
- If the search engine you are using offers guidelines for searching, be sure to follow its instructions.

Another search option is to check online encyclopedias. Although encyclopedias typically provide somewhat general information, many also offer links to other Web sites for the topic. These sites often contain specific information.

Using search engines effectively will enable your students to find information on the Web that can support them in their work on math projects. Understanding how to access information the Internet contains is a valuable skill that extends well beyond math class.

Conclusion

A math class in which students are actively engaged in working on projects appears on the surface to be quite different from a traditional math class. However, a close look shows that these seemingly different models have much in common. In both, students are learning math, discipline is necessary, and motivation is crucial. In the traditional math class, however, students sometimes fail to recognize the far-reaching importance mathematics has in our lives. They do not realize that math is just about everywhere. Math projects demonstrate to students that it is. Projects not only show the connections of math to other subjects but offer students the chance to incorporate various skills, strategies, and methods in finding solutions to meaningful problems.

Hands-on math projects with real-life applications

Chapter Three

Assessing math projects

As the objectives and methods of math teaching change, forms of assessment must change too. It is through assessment that we as teachers can validate the effectiveness of our instruction and evaluate our students' understanding of mathematics. Assessment should always be viewed as an essential part of the learning process and should involve both you and your students.

You may choose from a variety of tools when assessing your students' work on projects, including observation logs, checklists, and point systems. Because these assessment methods are flexible, you can tailor them to meet your needs. Combining the assessments of projects with the tests and quizzes of the general curriculum can give you a detailed profile of your students' overall achievement in your class.

Observation Logs

As you move around the classroom during project work, you may observe students working individually or in groups. Writing down your observations will provide you with a permanent record of their

INDIVIDUAL OBSERVATION LOG

Name _____ Section _____

Project _____

Date	Comments

GROUP OBSERVATION LOG

Names _____

_____

_____ Section _____

Project _____

Date	Comments

progress. A practical way to do this is to use either the "Individual Observation Log" or the "Group Observation Log," each of which is provided.

To reduce your workload to a manageable level, plan to observe only five to ten students per day in each class. Attach individual log sheets to a clipboard, and carry it with you around the room, focusing your attention on the students you wish to observe that day.

When completing the logs, you may record items that indicate mathematical thinking, understanding of concepts, insights, or reflections. You might also note behavior. Selecting two or three skills or behaviors to concentrate on reduces the chances you will feel overwhelmed with things to look for. It is helpful to develop your own system of shorthand using abbreviations, codes, and phrases—for example:

- Identify names with initials. John becomes *J*.
- Abbreviate frequently used words. Excellent is *ex*, good is *g*, fair is *f*, well is *w*, strategy is *strat*, work is *wk*, process is *proc*, question is *quest*, group is *gp*, illustrate is *il*, problem is *prob*, needs improvement is *ni*.
- Use phrases rather than full sentences whenever possible.

Here is a sample entry on an individual log: "Wked w with gp. Offered sketch to il prob."

Conferences provide a fine opportunity to gain an understanding of your students' growth in mathematics. Simply talking to students about the project they are working on can give you insight into their thoughts and feelings about math.

While you can learn much about your students when they ask you questions, you can also pose specific questions to your students that will help show their understanding of math. Such questions may focus on comprehension of problems, formulation of strategies, procedures, calculations, justification of solutions, relationships between ideas, or group cooperation. Having a list of questions prepared ahead of time can help you zero in on points you wish to address. Since many of the students in a class frequently share the same problems and concerns, asking a few students the same questions will often provide information about the class's general thinking. See "Possible Assessment Questions" for a list of questions that you can ask during observations and conferences.

Checklists

Checklists are another useful tool for observation. Unlike an observation log in which you write notes detailing the progress of students, a checklist is an assortment of predetermined skills and behaviors. As you observe one of the skills or behaviors on your list for a particular student, you simply check it off. As with the observation log, it is practical to select five to ten students per day in each class on whom to focus your attention. While a checklist may include numerous

Possible Assessment Questions

Asking students questions about their work can provide you with valuable insight about their progress. The following questions are just some of the possibilities:

- What is this problem asking? How would you explain it to a friend?

- What must you find before you can come up with a solution?

- How are the facts of this problem connected? How does one fact relate to another?

- Is there any information in this problem that you do not need? What is it, and why is it not needed?

- Is there any information missing in this problem that is necessary to solving it? How would you go about finding it?

- What strategies might you try to solve this problem? Which do you think is the best one? Why?

- Can the information or facts presented in this problem be arranged in a pattern? In what way? How might that pattern help you solve the problem?

- Would drawing or sketching help you to solve the problem? If yes, how?

- How might you share your understanding of the problem with your group?

- How might your group divide tasks in solving this problem?

- How might your group work more effectively?

- What is the best solution to the problem?

- How can you justify your solution?

SKILLS CHECKLIST

Name _____ Section _____

Project _____

E = Exceptional   S = Satisfactory   N = Needs Improvement

Skill	Date						
Defines problem							
Identifies useful strategies							
Implements strategies							
Eliminates unnecessary data							
Collects needed data							
Organizes data							
Analyzes and interprets data							
Finds relationships							
Uses models							
Tests strategies							
Demonstrates solutions							
Explains results orally							
Explains results in writing							
Uses logic in arguments							
Makes estimates							
Makes accurate calculations							
Uses technology							
Cooperates with group							
Supports group members							
Shares ideas with others							
Listens to others' ideas							
Remains on task							
Is persistent							
Demonstrates creativity							
Shows enthusiasm							
Tries new ideas							
Takes risks							
Is confident							

Comments:

skills, you may decide to concentrate on only a few, selecting those that apply best to particular projects.

A sample "Skills Checklist" is included for your use. You may use it in its current form or as a reference for designing your own. The checklist presented here is set up for seven days. It also provides space for comments should you wish to record something in more detail.

Point Systems for Project Assessment

Some teachers prefer, and many schools require, numerical scores or grades for the work students do. Since most projects are long-term, complex activities, it is not easy, or usually fair, to give students a grade based simply on completion. A system where points are assigned to specific parts of the project is an alternative.

While you can break down point totals to fit your personal grading criteria, the point system that is included in this section works well for most projects. It is based on a total of 100 points, which can be easily translated to percentages. See "Grading Projects Using a Point System."

Evaluating Writing

Many math teachers feel uncomfortable evaluating the writing of their students. However, with the growing understanding that writing should be an integral part of a math curriculum and that it often has a major role in math projects, more and more math teachers will be reading and commenting on the writing of their students. When your students write about mathematics, they share with you much of their understanding of and attitudes toward math.

You may evaluate various types of student writing, including articles, essays, reports, solved problems, biographies of mathematicians, written logs of problem solving, or simple ponderings over tough problems. The exception here is math journals; we recommend that you do not grade journals. Journals (see Project 31) are storehouses of a student's thoughts, reflections, and impressions about math. Once you start grading them, many students will begin writing what they think you want to see, translated to, "I'll get a better grade this way." Once that happens, you will no longer find the honesty that can be so valuable and refreshing in journals.

When you do grade the writing of your students, select a few points or criteria to focus on. This will make it easier to keep your objectivity. Also, rather than taking a pile of papers home each night to read, take only a workable number. If you try to do too much, you will become frustrated and probably lose perspective. Concentrate your evaluative efforts on content. Since writers enjoy responses from readers, offer comments to your students, addressing mathematical ideas and issues. Keep your comments upbeat and positive. When criticism is necessary, be sure it contains suggestions for improvement. For more information about grading writing, see "Suggestions for Grading Writing in Math Class."

Grading Projects Using a Point System

Following is an example of how a project's parts can be broken down and quantified. The total number of points is 100. You may use this system or design one of your own.

Outcome/Action/Behavior	Points
Satisfactory solution The solution is valid and practical.	25
Justification of results Students justified results through an oral presentation, written report, or discussion. They backed up the results with sound arguments.	15
Methods Students eliminated impractical procedures and focused their efforts on the most useful. If necessary, the students eliminated and found data. They were able to analyze and organize information and use technology where applicable.	15
Accuracy Reasoning and computation were logical and accurate.	15
Creativity Students showed original or insightful thinking.	10
Persistence Students did not give up.	10
Cooperation with group Students worked well together, shared ideas, and listened to the ideas of each other. They showed a willingness to help each other.	10

Suggestions for Grading Writing in Math Class

Grading the writing of students in math can be new territory for many teachers. The following ideas can help.

- Focus evaluation on content rather than mechanics.

- Offer comments and responses directly on student papers whenever possible.

- Keep comments positive. Offer specific suggestions for improvement.

- Comment on only one or two points. Mentioning more may only confuse or discourage students.

- Encourage students to edit each other's writing and revise their work before handing it in.

- Work with your students' English teacher in promoting effective writing techniques.

- Discuss with your students what you will be looking for during evaluation.

Following is a model for scoring student writing, based on percentages:

- *Focus:* The topic is clearly defined. All ideas support the topic. 20%

- *Content:* The student uses fresh, insightful, or original ideas. The topic is developed and supported with details. Mathematical reasoning is sound and shows an understanding of concepts. 25%

- *Organization:* The piece progresses logically from beginning to end. An introduction, body, and conclusion can be identified. 25%

- *Style:* The writing is appropriate for the topic and audience. Ideas are communicated effectively. There is a distinct voice. 15%

- *Mechanics:* The writer uses correct punctuation, grammar, and spelling. 15%

Self-Assessment

A successful math class in which learning is vigorously pursued is a place of continuous assessment on the parts of both teachers and students. While virtually all students expect their learning to be evaluated by their teachers, few have ever been asked to assess themselves. Self-assessment is perhaps the most valuable of any form of evaluation.

You should encourage your students to assess themselves. A good place to record thoughts about personal growth in math is in math journals. At the end of a project, ask your students to write a journal entry about the project. Suggest that they include the strategies they used, problems they encountered, and what they learned from the project.

If you prefer, you may distribute copies of the "Student Self-Assessment." Having students answer the questions on the assessment will help them to evaluate their own work and learning. (See Project 60 for a comprehensive assessment that may be applied to the entire year.)

While we as teachers continuously evaluate the efforts of our students, it is helpful for us to step back occasionally and assess ourselves. This is particularly true for teachers who are implementing projects for the first time. At the very least, you should assess yourself by considering what went well and what you would do differently next time. Asking yourself the questions contained in the "Teacher Self-Assessment" can be most helpful.

Conclusion

Assessment clearly is an essential part of the program of any classroom, with the purpose of promoting and assisting learning. It should be continuous and effective.

Hands-on math projects with real-life applications

Student Self-Assessment

Name _____ Date _____ Section _____

Project _____

To evaluate your work and what you have learned during this project, answer the following questions.

1. What did I like about this project? _____

_____

2. What did I not like about it? _____

_____

3. What strategies did I use to solve the problem? _____

_____

_____

4. Could I have used other strategies? If yes, which ones? _____

_____

_____

5. Did I justify my solution sufficiently? Could I have provided more proof? How?

_____

_____

_____

6. What did I learn during this project?! _____

_____

_____

_____

_____

Teacher Self-Assessment

Honest answers to the following questions can help ensure that your next project will be even more successful.

- Did I present the project clearly? If not, how might I make it clearer?

- Did the students understand what they were supposed to do? How might I help them understand better?

- Did I arrange the classroom appropriately? What could I change to make the classroom more conducive to project work?

- Were the students organized in effective groups? Who would I change?

- Did I monitor students effectively? Do I know what each student learned?

- Did I ask appropriate questions that provided guidance without giving away answers? What was my best question? What was my least effective?

- Did I provide enough time for sharing and discussion upon conclusion? If not, how might I arrange more time in my schedule?

- What would I do differently to improve this project?

Part Two

The Projects

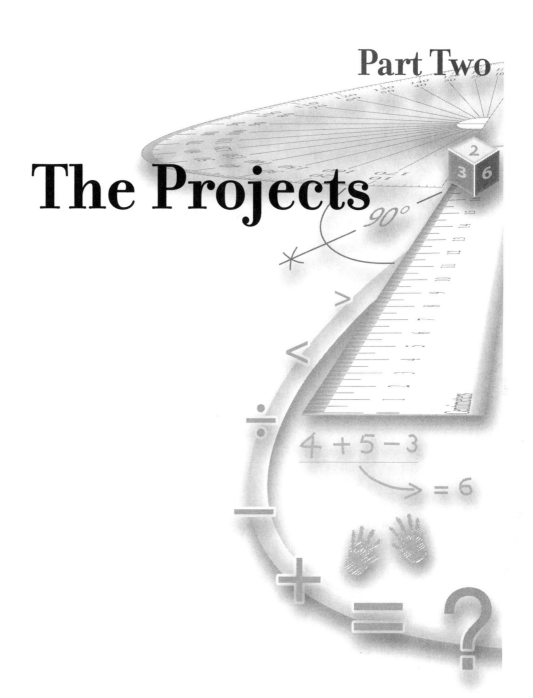

Math and Science

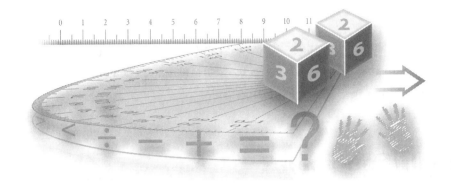

The benefits of recycling

Most communities across the country have recycling programs: newspapers, glass, aluminum cans, and plastic bottles are collected and transported to recycling centers, where they can be processed and used again in some form or another. While most people recognize the importance of recycling, few realize how significant the benefits of recycling truly are.

Goal

Working in groups of four or five, students will research recycling to determine who or what benefits from recycling programs. Each group will write a summary of its conclusions and present its findings to the class. *Suggested time:* Four to five class periods over two weeks.

Math Skills to Highlight

1. Using data as a basis for making decisions about recycling
2. Analyzing and simplifying numbers to choose an appropriate scale for a graph

3. Reading and creating graphs, tables, and charts
4. Using math and writing as a means to communicate ideas
5. Using technology in problem solving

Special Materials/Equipment

Reference books and articles about recycling; markers; felt-tipped pens; poster paper; rulers; compasses; protractors; scissors; paste; tape. *Optional:* Computers and printers to produce spreadsheets, graphs, and tables; Internet access for research; overhead projectors and transparencies for presentations.

Development

Before beginning this project, talk about recycling with your students. Most communities have recycling programs, so your students will be familiar with them. Most people assume that recycling is beneficial. But who actually benefits and in what ways? At the very least, recycling programs reduce stress on the environment by cutting the demand for natural resources. In addition, they create jobs (recycling companies have become thriving businesses), save money, and spur the development of new technologies.

- For this project, focus your groups on the main areas of recycling: newspapers, glass, aluminum cans, and plastic bottles. Although some communities also recycle grass clippings, electronics, and used motor oil, information on the first four will be easier to find.

- Distribute copies of Student Guide 1.1, and review the information with your students. Be sure they understand what they are to do. Point out that students should concentrate their research efforts not only on the benefits that recycling offers people but also its benefits for the environment.

- Mention that this project is quite broad and that groups may come to different conclusions. The groups should support their findings with facts.

- Hand out copies of Data Sheet 1.2, "Some Facts About Recycling." This sheet provides students with interesting information about recycling that may help them to get started with their research.

- Since students will need to conduct research, plan to spend at least two class periods in the library. A week or so ahead of time, inform your school's librarian about your project and ask her to reserve books on recycling.

- Encourage students to spend some of their own time researching.

- Encourage students to conduct research on the Internet.

48

- Remind students to keep an accurate list of their sources. This will make it easier to verify facts.
- Reserve at least one class period for students to meet in groups to analyze and organize their data. They should work toward a consensus regarding the benefits of recycling. To support their findings, suggest that they create posters, charts, graphs, tables, or transparencies for the overhead projector. You may set aside additional class time for them to work on their presentations.
- Suggest that students use computers to create spreadsheets, graphs, or tables to illustrate their results. Set aside at least one class period for the presentations. If you have an overhead projector in class, encourage students to incorporate its use in their presentations.

Wrap-Up

Students present their findings to the class.

Extensions

Visit a recycling center for a class trip. Invite a recycling expert to discuss recycling with the class or explore other areas of waste management such as nuclear waste.

STUDENT GUIDE 1.1

The Benefits of Recycling

Situation/Problem

Most communities across the country maintain recycling programs in which newspapers, aluminum cans, glass bottles, and plastic bottles are recycled. Most people assume that recycling is beneficial. But who actually benefits? For example, does recycling reduce stress on the environment? Does it improve the quality of life for people? Does it create new jobs, save money, or promote the development of new technology?

For this project, your group will attempt to answer the question, What are the benefits of recycling? You will write a summary of your conclusions and then present your findings to the class. Include graphs or illustrations that support your conclusions.

Possible Strategies

1. Divide the tasks of this project among group members. Tasks will include researching; analyzing and organizing information; creating materials such as graphs, posters, and other displays to support your conclusions; and presenting your conclusions to the class.

2. After gathering research, analyze your information and draw conclusions.

Special Considerations

- When organizing tasks, divide them equally. For example, one member may research newspaper recycling, another may research aluminum cans, and still others may research glass and plastic bottles. Some members may be responsible for creating posters to support your conclusions, while others may do the actual presentation.

The Benefits of Recycling *(Cont'd.)*

- During research, look for the benefits recycling offers to people, companies, and the environment. Support your findings with facts.

- Keep an accurate list of your sources by using a standard bibliographical format. Note specific page numbers where information was found. Your English text or a writer's stylebook has information on bibliographies. Also include any online or electronic sources you used in researching.

- After you have gathered all your information, analyze it, and draw conclusions about the benefits of recycling. Explain your conclusions in a summary. Your summary should contain three parts: (1) an opening in which you provide your results, (2) a body that offers details supporting your results, and (3) a closing in which you reemphasize the main point of your summary.

- Illustrate your results with graphs, charts, tables, or posters. Consider using computers to create graphs and tables from spreadsheets.

- Decide how you will present your findings to the class. Will one group member act as spokesperson, or will each member have a part in the presentation? Rehearse your presentation so that you are able to conduct it smoothly. Be ready to answer questions from the class.

To Be Submitted

1. Project summary
2. Notes and sources
3. Supplementary materials such as graphs, tables, or charts

DATA SHEET 1.2

Some Facts About Recycling

- Approximately 40% of the solid waste in the United States is paper and paper products.

- About 45% of the paper Americans use each year is recovered for recycling.

- Making 1 ton of recycled paper uses only 60% of the energy needed to make 1 ton of paper from fresh wood.

- One ton of paper made from recycled paper saves about 17 average-sized trees and 7,000 gallons of water.

- It takes more than a half-million trees each week to supply Americans with their Sunday newspapers.

- If each American recycled just 10% of the newspapers he or she reads, about 25 million trees would be saved each year.

- Recycling glass reduces energy costs for making new glass by close to 30%.

- Aluminum cans made from recycled aluminum saves up to 95% of the energy costs for making cans from new aluminum.

- Every minute in the United States, 113,000 aluminum cans are recycled.

- On average, each American uses 392 aluminum cans every year.

- About 63 billion (that's right!) aluminum cans are recycled every year in the United States.

- Every hour, 2.5 million plastic bottles are used in the United States.

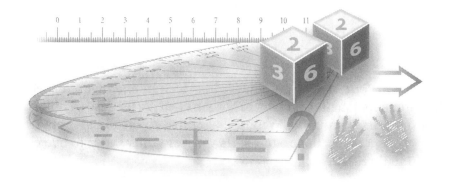

Endangered species: Can they be saved?

In 1973 Congress passed the Endangered Species Act to protect plant and animal species facing extinction. Any individual or organization may petition the Fish and Wildlife Service or the National Marine Fisheries Service to add a species to the Endangered Species List. Depending on its risk of extinction, a species may be listed as endangered or threatened. In either case, the species becomes protected by law. It is illegal to hunt, buy, or sell endangered species, their body parts, or products. Another component of the Endangered Species Act is to preserve ecosystems on which endangered plants and animals depend.

Saving endangered species is a complex issue that has become controversial as well. Some people believe that the costs do not justify the results, especially when preserving the habitats of endangered plants and animals comes into conflict with the loss of jobs or the curtailment of development. Other critics point out that for every success, there are many more failures. As they work through this project, your students will confront these and other issues that force a person to wonder: When a species stands in the way of human development, should it be saved?

Goal

Working in groups of four or five, students will select an endangered species and try to determine its long-term chances for survival. After researching and drawing conclusions, each group will report its findings to the class, supporting its views with posters, graphs, tables, or charts. *Suggested time:* Four to five class periods over two to three weeks.

Math Skills to Highlight

1. Researching, reading, and interpreting data
2. Using data to draw conclusions and make predictions
3. Creating graphs, tables, or charts
4. Using technology in problem solving

Special Materials/Equipment

Reference books and articles about endangered species; markers; felt-tipped pens; poster paper; rulers; protractors; compasses; scissors; paste; tape. *Optional:* Computers and printers to produce graphs and tables; Internet access for research; overhead projectors and transparences; PowerPoint or interactive whiteboards for presentations.

Development

You might consider collaborating with your students' science teacher on this project because of its focus on endangered plants and animals. Before introducing the project, discuss endangered species with your students. Ask them if they can name some endangered plants or animals. Ask if they know of species that were on the endangered species list once but have sufficiently recovered in population that they are no longer considered endangered. For example, the American bison, the American alligator, and the gray whale have all made significant comebacks. Unfortunately, despite protection, many species show slow improvement or even decline in number. Some examples are the California condor, the whooping crane, and the black-footed ferret.

- Begin this project by explaining to your students that each group will select a plant or animal currently on the Endangered Species List. They will research their subject, finding out why it is in danger of becoming extinct, as well as what is being done to save it. (Data Sheet 2.2, "Examples of Endangered Animals," lists some animals that face extinction.)

Based on their research, students will formulate a conclusion and make a prediction as to the chances of the species being saved.

- Distribute copies of Student Guide 2.1, and discuss the material with your students. In particular, draw their attention to the "Special Considerations" section for guidance about how they should conduct their research.

- Remind your students that any conclusions they draw must be supported by the evidence they uncover.

- You may wish to hand out copies of Data Sheet 2.2. This is only a partial list of endangered animals, but it is good for starting the project. Students can easily find comprehensive lists of endangered animals and plants in encyclopedias and other references, as well as on numerous Web sites.

- Distribute copies of Data Sheet 2.3, "Some Facts About Endangered Species." The information on the sheet can provide students with important background information as they begin their research.

- Encourage students to use computers for creating graphs, tables, and charts. Computers may also be used for researching online or accessing electronic databases.

- Plan on spending at least one class period in the library for research. Inform your school librarian a week or so in advance so that she can reserve materials on endangered species and be ready to help your students. Suggest to your students that the groups meet out of class if they need more time for research.

- Reserve at least one class period for groups to meet, analyze information, formulate conclusions, and begin to create materials to support their findings.

- At least one class period will be necessary for the presentations of the groups.

Wrap-Up

Each group presents its conclusions to the class. Consider incorporating PowerPoint or an interactive whiteboard in student presentations.

Extension

Have students write articles based on their research, and submit the articles to school, district, or PTA publications.

STUDENT GUIDE 2.1

Endangered Species: Can They Be Saved?

Situation/Problem

It has been estimated that over one hundred plants and animals are becoming extinct around the world each day. For this project, your group is to select an endangered species, research why it is in danger of becoming extinct, and what is being done to save it. Based on the information you find, you are to predict its chances for long-term survival. You will present your findings to the class.

Possible Strategies

1. Review a list of endangered species, and decide which animal or plant your group would like to research. You will be able to find lists of endangered species in encyclopedias and other references, as well as on numerous Web sites. (To find helpful Web sites, do a search using the term "endangered species.") To narrow your choices, do some preliminary research on several species.

2. Divide the tasks for this project among group members. You will need to be concerned with research, analysis of the facts, drawing conclusions, creating materials to support conclusions, and making your presentation.

Endangered Species:
Can They Be Saved? *(Cont'd.)*

Special Considerations

- In researching your species, concentrate your efforts on questions such as the following:

    Why is this species in danger of becoming extinct?

    Where is its habitat?

    Who or what is responsible for its being in danger?

    When have the greatest numbers of the species' population been lost?

    What is being done to save the species?

    What is the cost of the preservation efforts?

    Who is involved in trying to save the species?

    What are the chances of the species' long-term survival?

- Maintain an accurate list of your sources. Use a standard bibliographical format. Check your English book or a writer's stylebook for standard forms. Be sure to use a standard format when citing online sources.

- After gathering your information, analyze your facts, and draw conclusions. Include a prediction of the species' long-term chances for survival. Be sure to back up your predictions with facts.

- Support your conclusions with materials such as graphs, charts, and tables—for example:

    Stem-and-leaf plots

    Pictographs

    Bar graphs

    Line graphs

    Circle (or pie) graphs

- If possible, use computers for research as well as to help you create support materials.

- Plan your presentation. You may appoint one group member to share your findings and prediction, or each member may play a part. Rehearse your presentation so that it is smooth.

To Be Submitted

1. Notes and sources
2. Supplementary materials including any graphs, tables, and charts

Examples of Endangered Animals

The following animals have been identified as being endangered. This means that they are perilously close to extinction. Note that this is only a partial list. For a complete, updated list of endangered species, check current references in your library or online sources.

Endangered Animal	*Primary Habitat*
Big-horn sheep	California
Black-footed ferret	Western Canada to Texas
Blue whale	Oceans around the world
California condor	California
Cheetah	Africa, eastern Iran, India
Chimpanzee	Africa
Gorilla	Central and western Africa
Gray wolf	Eastern Europe, Russia, Canada, United States, Mexico
Florida panther	Florida
Humpback whale	Oceans around the world
Ivory-billed woodpecker	Southeastern United States, Cuba
Jaguar	Southwestern United States, northern Mexico, Central and South America
Ocelot	Southwestern United States, Mexico, Central and South America
Short-tailed albatross	Alaska, Oregon, Washington, Northern Pacific
Whooping crane	North America

DATA SHEET 2.3

Some Facts About Endangered Species

- Endangered species are animals and plants that are in danger of becoming extinct very soon unless they are protected.

- Passed in 1973, the United States Endangered Species Act makes it illegal to hunt, buy, or sell endangered species and their body parts or products. Another purpose of the act is to save ecosystems on which endangered plants and animals depend.

- In 1970, the rate of extinction around the world was estimated to be 1 species per day. By 1990, the rate of extinction was estimated to be 24 per day. Today, estimates range to over 100 per day.

- In the United States, over 1,000 plants and animals are listed as threatened or endangered.

- There are many natural causes of extinction, including floods, fires, diseases, volcanic eruptions, changes in climate, a loss of habitat or food required by the species, and calamities such as the catastrophic meteor hit that is believed to have caused the destruction of dinosaurs.

- Humans are the major cause of extinctions today through overhunting, overfishing, destruction of habitats, the overuse of pesticides, and the release of pollutants into the environment.

- Up to 20 million acres of rain forest are cleared each year for farming. Such destruction of habitats threatens countless plants and animals.

- Some 30 million species of plants and animals live in the earth's rain forests. This is half of all the plants and animals in the world.

- By the year 2025, if present trends continue, it has been estimated that at least 2 million species will become extinct because of rain forest destruction.

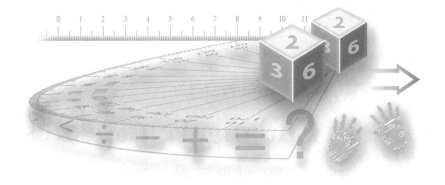

Charting your calories

Our bodies burn the calories obtained from food for energy. Since our bodies store extra calories as fat, calorie counting often becomes a major activity for people trying to lose or gain weight. In your class, for example, you might have a dancer trying to lose a few pounds, a football player trying to gain, and a wrestler concerned about maintaining his weight so that he may compete in a particular weight category. For many people, becoming aware of their caloric intake is an important part of understanding the body's overall nutritional needs. This project offers students the chance to trace the calories they consume. As they do, they will undoubtedly learn quite a bit about nutrition.

You might like to follow this project with Project 4, "How Many Calories Do You Burn Each Day?" Together, the two projects will help students realize the important role calories play in a healthy diet. Project 8, "The School's New Lunch Program," also focuses on nutrition.

Goal

Working individually, students will maintain records of the caloric content of foods they eat over a seven-day period. Upon completion of this period, they will average the calories they have consumed and write a brief report describing their daily caloric intake. *Suggested time:* One to two class periods. Much of the work for this project will be completed out of class.

Math Skills to Highlight

1. Reading charts and labels to find information
2. Collecting and organizing data
3. Using estimation and decimal computation
4. Determining the mean
5. Analyzing data to draw conclusions
6. Using writing and math as a way to share ideas
7. Using technology in problem solving

Special Materials/Equipment

Reference books and articles that contain the number of calories found in specific foods; assorted boxes and cans to show examples of food labels, which you may collect in advance. Students will also need to check the labels on the packages, cans, and bottles of the foods they eat to find the number of calories contained in the food. *Optional:* Computers and printers for writing reports; Internet access for research.

Development

Consider working on this project with your students' health instructor or the school nurse, who can offer valuable insights on nutrition. Introduce the project, and explain that a calorie is a unit of energy that a food supplies. Since for most people excess calories are stored as fat, balancing caloric intake with the body's caloric needs is a good way for an individual to attain his or her ideal weight.

- Start this project by telling students that they will chart the number of calories they eat each day for seven days. This information will help them realize their caloric needs and make them aware of the caloric payload of many foods.

- Distribute copies of Student Guide 3.1, and review the information with your students. Make sure that they understand where they can find the number of calories in the foods they eat. Also be sure they understand that the calories listed on package labels refer to serving amounts. A 16-ounce package of pasta, for example, might contain eight servings. Point out the recommended calorie consumption averages for young people noted on Student Guide 3.1.

- Obtain labels from various packages ahead of time, bring them to class, and show students how to read them. Focus their attention on the calories contained in the food or beverage and the serving size.

- Since students will be eating many foods for which they will not be able to check labels—for example, unpackaged foods at home, restaurants, and the school lunch—they will need to consult reference books and articles to find estimates of the calories these foods contain. Emphasize, however, that the best way to maintain an accurate account of calories is to refer to the label that comes with food. (For school lunches, you may wish to check with your cafeteria service, and find out if they can provide caloric information on the foods they serve to your students.)

- If you wish, you may hand out copies of Worksheet 3.2. Students may use this sheet to keep track of the calories in the foods they eat at each meal and snacks. You will need to photocopy seven sheets per student. Stapling them together helps prevent individual sheets from getting lost. Of course, you may instruct students to design their own sheets.

- Remind students to keep track of all the foods they eat for seven days. While it is best to record the foods for seven days in succession, if students miss a day, they may simply continue with the next. When you assign the project, consider giving them some leeway on the deadline. Making the project due in ten days gives students extra time if they need it.

- Encourage your students to consult references on the caloric content of foods on their own, using both print and online sources. We suggest that you also plan to spend a period in the library for students who will require extra time or help. Inform your librarian of this project so that she may reserve materials in advance.

- After charting their caloric intake for seven days, students are to find their daily average of caloric consumption. If necessary, review the steps for finding an average (mean). Students are to write a brief summary of their findings.

Wrap-Up

Collect the summaries. You may ask for volunteers to share their summaries; however, be aware that some students may be sensitive about sharing information about the calories they consume.

Extension

Have students reconsider their charts, and make recommendations for how they might improve their diets.

STUDENT GUIDE 3.1

Charting Your Calories

Situation/Problem

Many people pay close attention to the number of calories they consume each day. Some may be concerned with losing or gaining weight, while others simply wish to keep track of what they eat. For this project, you will record the number of calories you consume over a one-week period (seven days). At the end of the period, you will find your average caloric consumption, and write a brief summary of your findings.

Possible Strategies

1. Use a chart to record the calories you eat each day.

2. Be as accurate as possible in your record keeping. Although in some cases you may need to estimate calories, the more exact you are, the more your final totals will reflect your actual caloric intake.

Special Considerations

- Record the number of calories contained in each food you eat at every meal.

- If possible, record the calories at the end of the meal. If you wait too long afterward, you might forget. If you cannot keep a chart with you at meals, use a small notepad.

- Record the calories of all snacks.

- Water has no caloric content; virtually all other beverages do.

- Check the labels on packages, cans, or bottles for the number of calories a food or drink contains. This number usually appears as the number of calories "per serving." Be careful here, because most packages contain several servings. Multiply the calories by the number of servings you actually eat.

- Be sure to include the extras, such as butter, salad dressings, and mayonnaise. Adding butter to a slice of toast, for example, increases the total number of calories.

63

Charting Your Calories *(Cont'd.)*

- If you eat at a restaurant, politely ask the server if he or she can tell you the number of calories in the food you have ordered. Some restaurants have such information and gladly provide it to their patrons. If not, simply record the foods you eat, and consult a reference book or online source later to estimate the caloric values.
- After recording the calories you consumed each day, find your average daily caloric intake over the seven-day period.
- Compare your average to the following U.S. government recommended average calorie intake amounts. The amounts are shown in ranges. The more active a person is, the more calories he or she requires each day.

    Females, ages 9–13, 1,600–2,200 calories per day

    Females, ages 14–18, 1,800–2,400 calories per day

    Males, ages 9–13, 1,800–2,600 calories per day

    Males, ages 14–18, 2,200–3,200 calories per day
- For your summary report, share your findings and thoughts. Remember to use an opening, a body with supporting details, and a closing.

To Be Submitted

1. Chart of caloric intake
2. Summary report

Notes

_____

_____

_____

_____

_____

_____

_____

_____

_____

Name _____

A Daily Calorie Chart

Day Number _____   Date _____

	Type of Food	# of Servings	×	# of calories per serving	=
B r e a k f a s t					
		Total breakfast calories =			
L u n c h					
		Total lunch calories =			
D i n n e r					
		Total dinner calories =			
S n a c k s					
		Total snack calories =			
		Total for Day =			

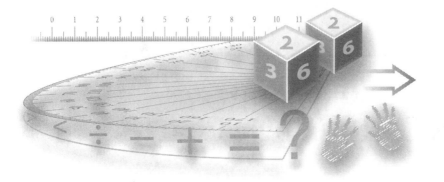

How many calories do you burn each day?

The calories our bodies obtain from food are either stored as fat or burned for energy. The more physically active a person is, the more calories he or she will use. You can expect to use far more calories running a marathon than if you watched a marathon on TV while stretched out on the couch. This project will help your students to understand the relationship between calories and physical activity.

Consider using this project as a follow-up for Project 3, "Charting Your Calories."

Goal

Working individually, students will record their physical activities throughout the day for a seven-day period. They will tally their daily caloric expenditures and find their average caloric expenditures. They will then summarize their results in a written report. *Suggested time:* Two to three class periods.

Math Skills to Highlight

1. Collecting and organizing data
2. Using estimation and decimal computation
3. Calculating with units of time
4. Finding the mean
5. Using writing and math as a way to share ideas
6. Using technology in problem solving

Special Materials/Equipment

Reference books that contain the numbers of calories used during specific physical activities; calculators. *Optional:* A scale for students to weigh themselves (however, keep in mind that some students might be sensitive about their weight; do not require that students weigh themselves in school); computers for Internet access for research.

Development

Consider working with your students' physical education or health teacher for this project. They will undoubtedly be able to offer advice, suggestions, and information. Before beginning the project, explain to your students that our bodies use the calories in food for fuel. Our bodies burn (or use through chemical change) calories for energy. Excess calories are stored as fat.

- Start the project by telling your students that they will be required to maintain records of their physical activities for the next seven days. They will then find a caloric value for each activity. Multiplying this by their weight and the time they spend at the activity will give them the approximate total number of calories they used.

- Hand out copies of Student Guide 4.1, and review the information with your students. Emphasize that they are to maintain a record of all the physical activities they do each day for seven days.

- Distribute copies of Data Sheet 4.2, "Caloric Expenditure and Physical Activities." Review the material with your students. Make sure that they understand how to use the formula for finding the total number of calories burned for specific activities. To use the formula, students will need to know their weight. You might provide a scale in class or suggest that students weigh themselves at home or in the gym. *A word of caution here:* Be sympathetic to the feelings of students who might be sensitive about their weight. Allow these students the option of weighing themselves outside class.

- Point out that the caloric values on the data sheet are approximations. Although different sources give slightly different values, most fall within the same range.

- Emphasize that although the data sheet provides a variety of activities, it is likely that some activities will not be listed. In such cases students should consult exercise or physical activity books in the school or public library or online sources. Since the data sheet provides so many activities, research time should be minimal. Encourage students to consult references on their own time.

- If you wish, you may distribute copies of Worksheet 4.3, which students can use to chart their activities and caloric expenditures. If you decide to use this worksheet, you will need to photocopy at least seven pages per student. Stapling them together reduces the chances that students will lose individual pages. Make additional sheets available for students who need more. If you prefer, suggest that students design their own charts.

- Although students will monitor their activities for seven days, you might want to give a nine- or ten-day deadline for completion of the charts. If students miss a day, they can make it up the next day.

- On completion of their charts, students are to find their total caloric expenditures for each day and their daily average caloric expenditures.

- When they are done, they are to write a report, summarizing their findings.

Wrap-Up

Discuss the relationship between calories and physical activities. Once again, keep in mind that some students might be sensitive about their weight and caloric intake.

Extension

Invite a personal fitness trainer to speak to your class about the importance of exercise.

STUDENT GUIDE 4.1

How Many Calories Do You Burn Each Day?

Situation/Problem

Physical activity burns calories. But do you know how many calories you burn each day? This project will help you find out. You will keep track of your activities for seven days. Based on the time you spend in each activity, you will calculate the total number of calories you used for the activity. You will then find the total number of calories you expended each day.

Possible Strategies

1. Use a chart to record the activities you perform each day.
2. Be as accurate as you can in recording activities and times.

Special Considerations

- Record every activity you do from the time you wake up in the morning to the time you go to sleep. Also be sure to record your sleep time.

How Many Calories Do You Burn Each Day? *(Cont'd.)*

- Along with each activity, record the length of time you were involved with it. Convert times to decimal equivalents based on 1 hour. For example, a half-hour would be 0.5, and 15 minutes would be 0.25.

- Try to record activities as you do them. If this is impossible, at the end of the day, review all the things you did and write them down on your chart. It is important to record each activity.

- Use Data Sheet 4.2 to find values for caloric expenditures. For activities not on this sheet, consult reference books or online sources. Even then, you may need to estimate some activities.

- If you must estimate the calories used during some activities, select similar activities, and base your estimations on them.

- To use the formula on Data Sheet 4.2, you will need to know your weight. If you are not sure and do not have access to a scale, use an estimate.

- Use a calculator to find the total number of calories used during specific activities.

- Total all the calories spent on all the activities for each day.

- Analyze your results and consider the following questions:

  Are you more active during the week or on weekends?

  What activities do you expend the most calories on? The least?

  Do you think your average caloric expenditure will be about the same throughout the year, or do you think it will vary? Explain your answer.

  Did your results surprise you in any way? Explain.

- Write a report summarizing your findings. Be sure to write clearly, use an opening, a body with supporting details, and a conclusion.

To Be Submitted

1. Chart of physical activities and caloric expenditure
2. Summary report

Caloric Expenditure and Physical Activities

Below are various activities and estimates of the amount of calories you would burn each hour for each pound you weigh while taking part in an activity. You can find an estimate of your caloric expenditure by using this formula:

Your weight × calories per hour per pound × time = Total calories

Suppose you weigh 120 pounds and mow the lawn for an hour and a half. You would multiply 120 × 2.7 × 1.5, which equals 486 calories. By mowing the lawn for an hour and a half, you would have used 486 calories, roughly equal to that hamburger and French fries you gulped down for dinner.

In the following list, the number following the activity is the calories per hour per pound you would burn during the activity.

Badminton—2.7

Baseball—2.9

Basketball—4.5

Boxing—4.5

Canoeing (leisurely)—1.2

Card playing—0.7

Chopping wood (ax)—2.3

Cleaning (house)—1.6

Cooking—1.3

Cycling—2.5

Dancing (ballroom)—1.6

Dancing (current hits)—2.8

Eating—0.8

Fishing—1.7

Football—4.4

Gardening—2.1

Golf (walking)—2.3

Gymnastics—3.7

Hiking—3.6

Horseback riding—2.7

Ironing—0.9

Jogging (distance)—4.2

Judo (vigorous) 4.3

Jumping rope—3.8

Keyboarding—0.8

Lying at ease—0.6

Mowing the lawn—2.7

Marching (rapid)—3.9

Playing drums—1.8

Playing flute—1.0

Playing piano—1.1

Playing trumpet—0.9

Playing violin—1.3

Racquetball—4.0

Raking leaves—2.3

Rowing machine—3.1

Shoveling snow—3.9

Sitting—0.6

Skating—2.8

Skiing (cross-country)—3.7

Skiing (downhill)—2.5

Sleeping—0.4

Soccer—3.7

Swimming—3.8

Tennis—2.5

Walking—2.2

Weight training—1.9

Writing—0.8

Name _____

A Daily Activity Chart

Day Number _____   Date _____

Starting Time	Ending Time	Length of time	Activity	Cal. per hr. per lb.	Weight	Total Calories
				Total Calories for Day		

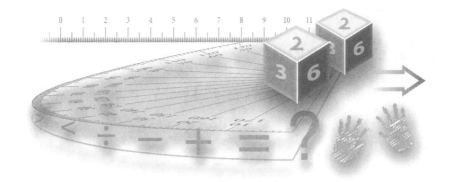

It is only natural

We hear the phrase, "Math is all around us," all the time. We often say it to students, most of whom nod their heads in response but in fact do not see much math at all. This is not because it is not there; they simply do not know where to look. This project will point them in the right direction.

Goal

Working in groups of two or three, students will make a collage by cutting pictures out of magazines or drawing them to illustrate examples of mathematics in nature. *Suggested time:* Two class periods.

Math Skills to Highlight

1. Identifying two- or three-dimensional shapes
2. Identifying types of symmetry as they appear in nature
3. Recognizing instances of mathematics in the natural world
4. Identifying instances of the Fibonacci sequence (optional)

Special Materials/Equipment

Old magazines; scissors; poster paper; glue; transparent tape; rulers; crayons; markers; colored pencils.

Development

This project is an excellent conclusion to a unit on geometry because many of the examples students will find of mathematics in the natural world will be geometrical. Since an important resource for this project is old magazines, start collecting them a few weeks in advance. Ask students and colleagues to bring some in. Your school's librarian might be able to help too. If possible, keep cardboard boxes in your classroom into which people can drop magazines.

- Explain to your students that they will work in pairs or groups of three to create a collage that shows examples of math in the natural world. They are to find these examples in old magazines or newspapers or draw them.

- Show students some examples of math in nature. Your textbook and reference books in the library will contain examples. A snowflake, which has six sides, is a hexagon. The sun, moon, and planets are spheres. Our galaxy, the Milky Way, is a spiral, as are some kinds of snail shells.

- If you have already studied symmetry with your class, ask your students to identify symmetry in some of the things they find.

- Distribute copies of Student Guide 5.1, and review the information with your students. Make sure they understand what they are to do.

- Hand out copies of Data Sheet 5.2, "Math in Nature." This sheet offers examples in nature where math is apparent and easy to find. This list will help your students in their search for material for their collages.

- Depending on the level of your class, you may wish to discuss the Fibonacci sequence.

- Suggest to students that they try to develop a theme for their collages based on Data Sheet 5.2. Of course, they may combine several themes or even create a collage that sweeps across all of nature. Encourage them to be creative.

- In creating their collages, have students cut and paste or draw examples of math in nature. They should number each picture and on a separate sheet of paper identify the math it represents. They should attach their identification sheet to the side or bottom of their collages when they are done.

Wrap-Up

Display the collages in the class or hallway. Discuss the many examples of math students found in nature.

Extension

Suggest that students find examples of math in the artificial world, such as in architecture.

STUDENT GUIDE 5.1

It Is Only Natural

Situation/Problem

You and your partner or group will make a collage of examples of mathematics found in nature. You will cut out pictures from old magazines or newspapers or draw pictures of your own to create your collage.

Possible Strategies

1. Brainstorm with your partner or group to identify geometrical shapes and properties that you already know about. For example, you know that the earth is a sphere.
2. Try to develop your collage using a variety of items that illustrate different shapes and properties in math.

Special Considerations

- Scan old magazines and newspapers for pictures you can cut out to use on your collage. Cut out pictures carefully and neatly.
- Consider drawing some pictures.
- After collecting several pictures, place them on your poster paper, and arrange them in a visually pleasing way.
- Select a title for your collage.
- Identify each picture on your collage by a number. List the numbers on a separate sheet of paper and explain the geometrical shape or mathematical concept represented by the picture. Attach this sheet to your collage.
- Your collage should focus on the natural world. Avoid including any pictures of man-made objects.

To Be Submitted

Your collage and list of mathematical shapes or concepts.

76

Name _____

Math in Nature

Following are some examples of mathematics in nature. There are many more:

Pinecones—spirals and Fibonacci numbers

Sunflowers—spirals and Fibonacci numbers

Ponds and lakes—reflections

Barchan dunes—crescent-shaped sand dunes

Star sapphire—pentagon

Queen Anne's lace—symmetry

Butterfly—symmetry

Pineapple—spiral, Fibonacci numbers

Fruits (cross-sections)—symmetry

Fairy rings (mushrooms)—circles

Ripples on water—circles

Diatoms—circles

Growth rings (trees)—concentric circles

Milky Way galaxy—spiral

Constellations—angles

Spider webs—concentric polygons

Planets, moons, stars—spheres

Flower stalks—lines or arcs

Salt—cubic crystals

Tree trunks—lines or cylinders

Earthworms—cylinders

Orbits of planets—generally circular (with slight variations)

Orbits of comets—elliptical or parabolic

Snowflakes—hexagons

Chambered nautilus (snail)—spiral

Common starfish—pentagons

Honeycombs—hexagons

Many animals (including humans)—symmetry

Morning glory buds—spirals

Daisy heads—spirals

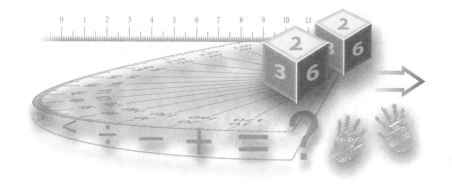

Designing a flower bed

The design of a flower bed—whether for an office building, park, school, or backyard—requires many skills involving math. Measurement, geometry, scale drawing, estimation, and computation of total costs make this project both interesting and comprehensive. On completion of the project, students will have gained a better understanding of plants and flower beds and may even wish to design a flower bed for their own homes.

Goal

Working in groups of three or four, students will be given a budget and a list of plants from which to choose. They are to design a flower bed and calculate what the total costs will be. *Suggested time:* Three to four class periods.

Math Skills to Highlight

1. Using geometrical shapes to design the flower bed, with either pencil and paper or computer software

2. Using a scale to make a diagram

3. Estimating costs and planning within a budget

4. Calculating with money, and making an itemized tally of expenses

5. Using technology in problem solving

Special Materials/Equipment

Rulers; reference books on plants; graph paper; measuring tapes; yardsticks or meter sticks; crayons or colored pencils should students wish to color their designs. *Optional:* Calculators, computers, and printers.

Development

With the emphasis of this project on plants, you might want to work with your students' science teacher.

Consider taking students outside to observe flower beds or gardens. This will sharpen their visual perception, stimulate their imaginations, and enhance their sense of relationships. You might also have students find the dimensions of a flower bed with measuring tapes or yardsticks to give them an impression of size.

- To start the project, distribute copies of Student Guide 6.1, and discuss the information. Explain that each group will work independently to create a unique design for a flower bed. Discuss the skills that students will likely use to complete the project, and encourage them to refer to this guide as often as needed.

- Hand out copies of Data Sheet 6.2, "Facts and Prices of Selected Plants." This sheet provides descriptions and costs of trees, shrubs, and flowers that students might use in their flower beds. The prices are general, and the plants on the sheet grow well in much of the continental United States. If you wish to provide a broader assortment of plants and prices, obtain and hand out the price lists and sales material from local nurseries and discount stores. If you do this, consider presenting the project in the spring, the time when garden centers bombard the public with sales material, which you can easily make available to your students, providing them with plenty of additional choices. You may also suggest that students consult reference books or online sources to learn about other plants they might use.

- Pass out copies of Worksheet 6.3, a grid on which students may sketch their design. Provide several sheets, and suggest that students try various arrangements. An option is for students to design their flower beds on computers. Various software packages, particularly drawing and art programs, enable students to create, arrange, and rearrange geometrical shapes. When working with designs on a grid, remind students of the importance of a scale. Perhaps each side of a block will be equal to 2 feet. Mention that students might wish to color their designs to obtain an accurate pictorial display of the plants they selected.

Wrap-Up

Make photocopies of each group's final designs, and share them with all the class members. A group spokesperson should be ready to explain and justify the group's selections.

Extensions

Design a real flower bed for the school or one for home. Invite a nursery representative to speak to the class about how flower beds are designed. Estimate the cost of maintaining the flower bed throughout the year by including the costs of fertilizer, water, weed and insect control, and possible replacement of plants that die.

STUDENT GUIDE 6.1

Designing a Flower Bed

Situation/Problem

Imagine that your school's PTA has donated $500 toward the design of a new flower bed in the school courtyard. Your class has been chosen to provide the design and select the plants. The area for the flower bed is 30 feet by 40 feet and enjoys full to moderate sunlight. Since the PTA would like to see several possible designs, your class will be divided into groups, with each group responsible for coming up with its own design. Your group will be given a list of possible plants from which to choose, including prices. Your goal is to design an attractive flower bed at the best cost.

Possible Strategies

1. Study some actual flower beds to gain an impression of size, color, and arrangement of the plants.
2. Sketch designs of possible flower beds on graph paper. Set up a scale. Place plants on the sketch where you think they would look best in the actual flower bed. You may prefer to design your flower bed on a computer, using software that has art and drawing components.

Special Considerations

- Visualize what your flower bed will look like now and in the future.
- Consult references to find out more about possible plants you might use.
- Seek a balance between the plants you select and the overall cost of the flower bed.
- Take into account the flowering periods of plants. Tulips, for example, bloom in the spring; marigolds bloom from summer to fall.

Designing a Flower Bed *(Cont'd.)*

- Consider how much sunlight the plants you select need. Also consider the amount of water necessary.

- Consider whether to choose deciduous or evergreen trees and shrubs. Deciduous plants lose their leaves in the fall; evergreens remain green throughout the year.

- Consider upkeep. Daffodils are perennials: they return each spring. Impatiens are annuals: they do not return each year and will need to be replanted. This will require new funds and labor each year for planting.

- Groundcover plants such as rug junipers do precisely what their name implies: they remain low and cover the ground. They are generally hardy and require little upkeep.

- Many plants, especially shrubs and trees, are sold in gallon containers that are related to the size of the plants. A plant in a 3-gallon container, for example, is smaller than one in a 5-gallon container.

- Generally the most attractive flower beds are those that have a mixture of flowers, shrubs, and trees with pleasing colors.

To Be Submitted

1. A final design of the flower bed, including a scale
2. Itemized tally and final cost of expenses

Notes

_____

_____

_____

_____

_____

_____

_____

_____

DATA SHEET 6.2

Facts and Prices of Selected Plants

Unless otherwise noted, the plants listed below are perennials. They generally tolerate varying amounts of sunlight and grow well in many different regions of the country. Heights given in the descriptions are adult sizes; heights given with prices are the size at the time of sale.

Trees

Japanese maple—red-leafed deciduous tree, 15 to 20 feet. Full sun to light shade. 4 feet high: $24.95.

Dogwood—deciduous, white or pink flowers in late spring, 20 to 30 feet. Half sun to light shade. White, 4 feet high: $19.95; pink, 5 feet high: $24.95.

American holly—slow-growing evergreen, 45 to 50 feet. Full sun to light shade. 4 feet high: $29.95.

Canadian hemlock—evergreen, graceful pyramidal shape, 40 to 70 feet. Full sun to light shade. 5 feet high: $24.95.

Shrubs

Azalea—variety of flower colors. Blooms in spring. Up to 4 feet. 3 gallon, 12 inches high: $5.95.

Rhododendron—evergreen up to 10 feet. Red, purple, and white are typical flower colors. Spring blooming. 3 gallon, 18 inches high: $7.95; 5 gallon, 24 inches high: $12.95.

Dwarf burning bush—deciduous, up to 6 feet. Brilliant red leaves in fall. 3 gallon, 12 inches high: $9.95.

Forsythia—fast growing. 8 to 10 feet, bright yellow flowers in early spring. 6 feet high: $9.95.

Common juniper—evergreen. 5 to 10 feet, spiny blue-green leaves. 5 gallon, 12 inches high: $8.95.

Blue rug juniper—evergreen. Excellent ground cover with blue-green foliage. 3 gallon: $3.95.

Facts and Prices of Selected Plants *(Cont'd.)*

Flowers

Daffodils—hardy bulbs. 4 to 18 inches. Variety of colors. Spring blooming. Pack of 20 bulbs: $4.95.

Tulips—hardy bulbs. 6 to 30 inches. Variety of colors. Spring blooming. Pack of 20 bulbs: $5.95.

Common petunias—10 to 18 inches. Variety of colors. Bloom from late spring to the first frost. Annuals except in the mildest of areas. $.89 each.

Geraniums—12 to 24 inches. Variety of colors. Bloom from late spring to the first frost. Annual. $3.95 each.

Marigolds—6 to 18 inches. Variety of colors. Bloom from summer to first frost. Annual. 6 for $1.99.

Impatiens—6 to 18 inches. Variety of colors. Bloom from summer to first frost. Excellent for shady areas. Annual. 6 for $1.79.

Hyacinths—hardy bulbs. 12 to 15 inches. Mostly white, blue, or purple. Spring blooming. Pack of 20 bulbs: $6.95.

Lilies—hardy bulbs. Up to 36 inches. Variety of colors. Summer to early fall blooming. Pack of 12 bulbs: $8.95.

Chrysanthemums—1 to 4 feet. Variety of colors. Late summer and fall blooming. $4.95 each.

Crocuses—hardy bulbs. 2 to 6 inches. Variety of colors. Late winter blooming. Pack of 20 bulbs: $7.95.

Mulch is available at a cost of $3.95 per 2 cubic-foot bag, which will cover 16 square feet at a depth of one and a half inches. Mulch is decorative and helps retain soil moisture.

WORKSHEET 6.3

Centimeter Grid

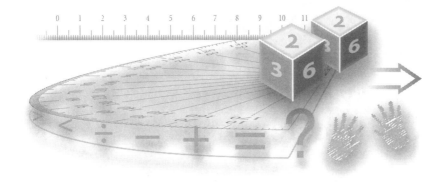

Buying a class aquarium

One of the major overall objectives of any math class is to prepare students for coping with the real world. At its simplest, this aim includes educating students so that they can evaluate and compare the features, benefits, and costs of items they wish to purchase. This project, which requires students to research and decide on equipment and materials necessary for setting up a class freshwater aquarium, will help them to acquire important real-life skills.

Goal

Working in groups of three or four, students will assume that they have been given $200 with which they may purchase a class aquarium. Their task is to buy the biggest tank, with its necessary accessories, and populate it with the greatest number of guppies or goldfish they can while staying within budget. *Suggested time:* Two to three class periods.

Math Skills to Highlight

1. Evaluating the features and benefits of items against their cost
2. Making decisions regarding price

3. Using charts and tables to obtain information

4. Tallying total costs

5. Using technology in problem solving

Special Materials/Equipment

Calculators. *Optional:* Catalogues of aquarium equipment and supplies; computers with Internet access for research.

Development

Explain to your students that aquariums are common in homes, businesses, and schools. People find aquariums to be interesting hobbies, and some researchers note that aquariums seem to provide people with a sense of relaxation and enjoyment.

- Start this project by telling students to imagine that the class has been given $200 by an anonymous donor (and obvious aquarium aficionado) to buy an aquarium for the classroom. Since many options are available in purchasing an aquarium, the class will be divided into groups, and each group will develop a proposal for the size of tank, accessories, and fish to buy. The proposals may not exceed the budget. Groups should attempt to select equipment and materials that use the money most wisely.

- Distribute copies of Student Guide 7.1, and review the information with your students. Be sure to emphasize the many factors detailed on the Student Guide that must be considered in buying an aquarium.

- Hand out copies of Data Sheet 7.2, "Aquariums and Accessories," which contains information on equipment, materials, and prices that students may use in developing their proposals. To keep the choices somewhat simple, we suggest that students select either guppies or goldfish for their tanks.

- If you wish to broaden the scope of the project, obtain and distribute copies of catalogues that contain information about aquarium equipment and accessories. Should you decide to do this, be sure to start accumulating the catalogues well in advance of the project so that you have enough for your students. Newspapers sometimes run advertisements of pet stores that sell aquariums and accessories.

- If your state has a sales tax, you may instruct your students to figure in its cost on the items they would buy.

- Remind students to select a spokesperson to present their proposal to the class. Their selections should be supported.

Wrap-Up

Groups share their proposals with the class.

Extensions

If possible, obtain money and set up a class aquarium based on the class's suggestions. Assume that the anonymous donor provided $75 more for fish. Checking catalogues, what other types of fish would students buy for their aquariums? Remind them to keep the size of the tank in mind and also the sensibilities of the fish (some fish do not get along well with others).

STUDENT GUIDE 7.1

Buying a Class Aquarium

Situation/Problem

An anonymous donor has given your class $200 to buy and set up a freshwater aquarium for the classroom. Since there are many different sizes of tanks and accessories, the class has been divided into groups. Each group will research aquariums and make a proposal for the size of tank, accessories, and fish to buy. You should select the biggest tank and most accessories you can but may not go over your budget.

Possible Strategies

1. Round off prices to do quick estimates of costs.
2. Select the tank, hood, light, and filter first. These will be some of your biggest costs.

Special Considerations

- Remember that fish need room to grow and thrive. Most aquarium supply shops suggest that you follow the rule of 1 inch for 1 gallon, that is, for every 1 gallon of water in your tank, you may have a fish 1 inch long. If you have a 10-gallon tank, for example, you may keep ten 1-inch fish or five 2-inch fish.

- While much equipment and supplies for aquariums are necessary, some are mostly decorative:

  Stones are colorful and provide a nice bottom but are not required.

  Water plants offer realism, allow fish hiding places, and add oxygen to water (supplementing the job of the air pump) but are not required.

  Snails and catfish help clean the aquarium but are not required.

- Use Data Sheet 7.2 for a list of items you may need and their prices.

- To reduce the chances that you will run over your budget, keep a running tally of the items you are considering.

Buying a Class Aquarium *(Cont'd.)*

- Pay close attention to costs. Some prices are given for tanks of a specific size.
- After you have decided on equipment, accessories, and fish, total your costs, and subtract them from your budget. Next, write down your proposal. Double-check your numbers, and be able to justify your decisions.
- Select a spokesperson to present your proposal to the class.

To Be Submitted

A copy of your proposal, including the items you would buy and the total costs.

Notes

_____

_____

_____

_____

_____

_____

_____

_____

_____

_____

_____

_____

_____

DATA SHEET 7.2

Aquariums and Accessories

Tank	Hood with Light	Filter
10 gallons $9.99	$19.95	$14.99
20 gallons $26.99	$29.95	$17.99
30 gallons $39.99	$34.95	$19.99

Air pump (suitable for all three tanks above), $14.99

Air line (8 feet long), $1.49

Air stone, $2.49

Charcoal, per 10-gallon tank, $5.19

Stones, 2-pound bag (about 1 pound per gallon is needed), $1.99

Fish food, 3.5 ounces, $8.99

Guppies (male or female), $3.25 each

Goldfish, $3.50 each

Snails, $1.99 each

Catfish, $3.49 each

Water plants, 6 to 8 inches, 4 for $2.29

Additional Notes

- Guppies are about 2 inches long.
- Goldfish may grow to be 5 to 7 inches, but will remain 3 to 4 inches in tanks where there is much competition among fish for food.
- Catfish are about 3 inches long.
- Filters are necessary to circulate and clean the aquarium's water.
- Charcoal is necessary for cleaning the water and reducing bacteria.
- Hoods act as a top for the aquarium (some fish try to "jump" out), and lights provide illumination.
- Heaters are needed for tropical fish but not for tanks that contain guppies or goldfish.

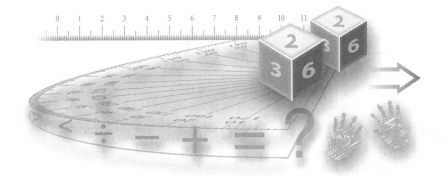

The school's new lunch program

Just about everybody in every school everywhere has a gripe about the school's lunches. No matter what the menu offers and no matter what efforts are taken in preparation, there are always some dissatisfied students (and teachers). Complaints range from the lack of choices to the lack of taste to high cost. While this project may not result in changing your school's lunch offerings, it will give students a chance to design potential menus that are tasty and nutritious.

Goal

Working in groups of four or five, students will design a five-day meal plan based on the major food groups. *Suggested time:* Three class periods.

Math Skills to Highlight

1. Classifying objects
2. Conducting a poll and using the results to make a decision

3. Obtaining information from food advertisements or labels

4. Estimating total costs

5. Using technology in problem solving

Special Materials/Equipment

Books on nutrition; circulars and advertisements from supermarkets; felt-tipped pens; markers; rulers; drawing paper for the menus. *Optional:* Computers and printers to design menus; an overhead projector and transparencies, PowerPoint, or an interactive whiteboard for presentations.

Development

Since it focuses on groups of food and making proper dietary selections, this project could be part of a science, nutrition, or health unit. Consider enlisting the aid of your school nurse or your students' science or health teachers.

- Start this project by explaining that students will have the opportunity to design a week's menus for the school's lunch program. Although they will be free to make any selections they want, they must follow dietary guidelines and keep the cost within the bounds the typical student would be willing to pay. (For most schools, caviar is out.)

- Distribute copies of Student Guide 8.1, and review the information with your class. Discuss how a poll is an excellent method by which to find out people's thoughts and opinions about an issue. In this case, polling students about their thoughts about lunch might result in uncovering fine ideas.

- Distribute copies of Data Sheet 8.2, "School Lunch Program." Explain that this sheet is a guideline detailing some types of foods that school lunches should contain. Your students should refer to it when designing their menus. Note that the requirements are minimums. Groups may design menus that exceed the minimum requirements.

- Point out that students may wish to consult nutrition books for foods that are not listed but may be substituted for foods on the list. They may also consult online sources.

- Remind students that as they are designing their menus, they must remain aware of the cost of the foods they are choosing. Suggest that they check the advertisements and circulars from supermarkets for current prices. They will need to calculate the total costs of each lunch for thirty people, then find the cost of an individual lunch for each person. They should price the lunches individually for each day.

- Hand out copies of Worksheet 8.3. Students may use this sheet for writing down their menus and computing costs.
- After they have decided on and priced their meals, they should design an attractive menu that includes the cost of each lunch. Computers that have software that has clip art and multiple fonts will enable students to create attractive menus, although they may simply use pens and markers to design their menus.

Wrap-Up

Each group shares its menu with the class. A spokesperson from each group presents the group's menu, detailing why certain foods were selected. Consider having students use an overhead projector, PowerPoint, or an interactive whiteboard in their presentations. After each group has shared its menu, conduct a discussion that examines the nutritional content of school lunches.

Extensions

Vote on the most popular menu. If students have access to a home economics or home living class, suggest that they prepare some of the foods on the menu.

94

STUDENT GUIDE 8.1

The School's New Lunch Program

Situation/Problem

Your group will create a five-day menu for your school's lunch program. Your menu should be both nutritious and popular with the students. You will be responsible for totaling the costs of the foods you select and setting the prices students will pay for lunch. Plan the lunch program for thirty students.

Possible Strategies

1. Poll your friends to find out what foods they would like the lunch program to offer.
2. Select foods from the major food groups.
3. Divide the tasks for this project.

Special Considerations

- Divide responsibilities according to the interests of group members. For example, two students may like to design and conduct the poll, two others may create the menu, and another may enter the menu on a computer. All members may work together to select the foods and find the prices.

- For a poll to be effective, the pollsters must decide on specific questions before contacting people. Try to think of three to five questions that you can ask others about school lunches. Avoid questions that may be answered with a simple yes or no, and include questions about foods and cost.

- Decide who you will poll. You may limit your poll to your class or ask students from other classes too. Remember that the more people you poll, the more accurate your information will be.

The School's New Lunch Program *(Cont'd.)*

- Refer to Data Sheet 8.2, which contains examples of foods from the major food groups. You may consult reference books or online sources for other types of foods to serve.

- In designing your menu, make sure that you meet basic nutritional requirements each day. (A lunch of chips and cola is *not* a healthy meal.)

- As you decide on foods for your menu, check advertisements and supermarket circulars for the prices of these foods. You may use Worksheet 8.3 to help you list foods and compute costs. Remember to buy enough food for thirty students.

- Do not calculate the costs of paper plates, napkins, or utensils.

- Make an attractive copy of your menu. Include the price you would charge for the lunch. Consider using computers to create and print your menu. You may also create an attractive menu using markers and pens.

- Present your menu to the class, explaining how and why you selected the foods you did. Be sure to offer support for your choices.

To Be Submitted

1. A copy of your menu
2. Your completed worksheet

Notes

_____

_____

_____

_____

_____

_____

_____

_____

Name _____

School Lunch Program

Your menu should offer lunches that provide at least one serving from each of the major food groups. Following are some examples from each group.

Meat/Poultry/Fish/Bean Group

Lunch meat
Hamburger
Hot dog
Tuna
Turkey
Chicken
Peanut butter
Eggs
Peanuts

Vegetable Group

Salads
Potatoes
Corn
Tomatoes
Broccoli
Carrots
Peas
Peppers

Bread/Cereal Group

Slices of bread
Rolls
Cakes
Pasta
Pizza
Rice
Tortillas
Noodles
Muffins
Bagels

Fruit Group

Apples
Bananas
Oranges
Pears
Grapes
Cantaloupe

Milk Group

Milk
Yogurt
Cheese
Cream cheese
Butter
Cottage cheese
Ice cream

Consult additional sources for more examples of foods.

A Five-Day Menu

	Menu	Need to Buy	Price
D A Y 1			
D A Y 2			
D A Y 3			
D A Y 4			
D A Y 5			

Total cost for Day 1 = _____ Cost per student _____

Total cost for Day 2 = _____ Cost per student _____

Total cost for Day 3 = _____ Cost per student _____

Total cost for Day 4 = _____ Cost per student _____

Total cost for Day 5 = _____ Cost per student _____

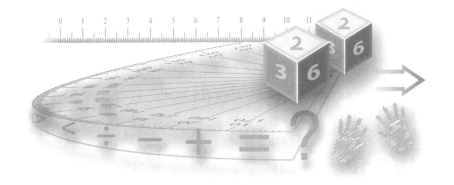

What is the weather?

Weather affects everybody. Because students have firsthand experience with it, the weather is a subject to which they can relate. It also provides an opportunity for students to practice their skills of observation, analysis, and interpretation.

Goal

Working in pairs or teams of three, students will observe and chart various weather conditions—including temperature, wind speed and direction, wind chill, barometer, precipitation, and cloud cover—for seven days. At the end of this period, they will organize and represent their observations in graphs, tables, or charts. They will also write a brief summary report of their observations. *Suggested time:* Two to three class periods.

Math Skills to Highlight

1. Using observation skills to collect data
2. Interpreting data

3. Expressing information in the forms of graphs, tables, or charts

4. Using writing as a means to express ideas in math

5. Using technology in problem solving

Special Materials/Equipment

Poster paper; rulers; felt-tipped pens; markers; colored pencils; daily newspapers (or access to TV and radio weather reports) for weather data. *Optional:* Computers and printers for creating graphs, tables, or charts, and writing summary reports. Computers with Internet access may also be used for accessing weather data online.

Development

Consider working with your students' science teacher for this project. Perhaps your students will be studying weather in science, and you may decide to assign the project to coincide with their science activities. To promote interest in the project, discuss with your students how weather plays a major role in our lives. Ask how the weather affects them. The softball game might be rained out; skiing might be poor because of the lack of snow; frigid weather forces people to wear heavy clothing; and hot, humid weather makes us seek air-conditioning.

- Start the project by explaining that students will work in groups of two or three. They will observe and chart various weather factors for seven days.

- Hand out copies of Student Guide 9.1. Review the information with your students. In particular, point out the weather factors under "Special Considerations" they are to observe and record.

- Emphasize that students are to record information about these weather factors for seven consecutive days. They may obtain their data from local newspapers, TV or radio weather reports, or online sources.

- Hand out copies of Data Sheet 9.2, "Weather Words." This sheet provides information about the weather. Suggest that students consult their science texts and other references for more information.

- After obtaining their information, students are to analyze and organize their data and then create posters that use graphs, tables, or charts to represent their data. They may represent their data in various ways. Give them some examples to avoid confusion. For example, bar graphs may be used to show the temperatures over the period. Double bar graphs may be used to show the daily high and low. Line graphs may be used to indicate wind chill. Some students may wish to simply display all of the data they find on a chart or table. You should provide suggestions based on the degree to which your students have studied and understand graphs.

The groups are also required to write a summary of their observations. You may wish to set aside one or two class periods for this work.

Wrap-Up

Display posters and reports. Conduct a class discussion about the observations.

Extensions

Obtain a copy of *The Farmer's Almanac* from last year, and compare the actual weather with what the almanac predicted. You may invite a meteorologist in to talk to the class about how mathematics is important to weather forecasting.

STUDENT GUIDE 9.1

What Is the Weather?

Situation/Problem

Working with your partner or group, you will chart specific weather conditions in your area for seven days. After compiling your data, you will organize and display your data, and write a report describing your observations.

Possible Strategies

1. To maintain the accuracy of your data, try to obtain information from the same source each day.

2. Divide the responsibility for obtaining data with your partner or group members. For example, you may collect the data on Monday, Wednesday, and Friday, while someone else takes the other days. Perhaps you will be responsible for finding the temperature and wind speed for all of the days, and your partners or other group members will find the other factors.

Special Considerations

- You are required to observe the weather, noting the following factors:

    Temperature—the high and low of the day

    Wind speed, average, and gusts

What Is the Weather? *(Cont'd.)*

Wind chill

Barometer

Degree of cloud cover (for example, partly sunny, mostly cloudy, sunny)

Precipitation

- In addition, you may record the following:

    Time of sunrise and sunset

    Normal high and normal low temperatures

    Coldest and highest temperature on each day in the past

    Severe weather (if any) through history on each day

- Accurately record your observations each day, and note your source. Consult local newspapers, watch TV weather reports, listen to radio weather reports, or consult online sources for information.

- Record, analyze, and organize your data. Use poster paper to create charts, tables, or graphs to represent your information. Be creative, add color, and use symbols to show weather details. For instance, illustrate a cloudy day by drawing a sky with plenty of clouds. Show a sunny day by drawing the sun. A partly cloudy day may be shown by drawing the sun with a cloud partially blocking it. You may even wish to draw a weather map.

- Consider using computers to create graphs, tables, or charts, as well as to write your summary report.

- Your summary report should describe your observations and sources. You should also explain why you illustrated your observations the way you did. Why, for example, would you select one type of graph over another? Write clearly.

To Be Submitted

1. Posters containing graphs, tables, or charts that represent observations
2. Summary report

DATA SHEET 9.2

Weather Words

Following are important words associated with weather and weather reporting:

Atmosphere   The thin envelope of gases, water vapor, dust, and other airborne particles that surround the earth. Commonly referred to as the "air."

Barometer   An instrument used to measure atmospheric pressure.

Blizzard   Heavy snow or blowing snow accompanied by winds of at least 35 miles per hour.

Climate   The average daily or seasonal weather of a particular place or region.

Cloud   A mass of small liquid water droplets or ice crystals suspended in the air.

Dew   Moisture that has condensed on objects near the ground.

Dew point   The temperature at which water vapor condenses to a liquid.

Drizzle   Precipitation consisting of very small droplets of water.

Fair   Weather condition in which clouds cover less than 40% of the sky, there are no extremes in temperature or wind, and there is no precipitation.

Fog   A cloud hovering close to the earth's surface.

Front   The area in which two air masses come into contact.

Frost   A thin layer of ice crystals that forms on the ground and other surfaces when water vapor condenses below 32 degrees Fahrenheit.

Gust   A sudden, brief increase in wind speed.

Hail   Precipitation consisting of pellets of ice.

High pressure   An air mass associated with fair weather.

Humidity   The amount of water vapor in the air.

Hurricane   An intense storm that generally forms over tropical regions. It brings heavy rain, strong winds, and powerful storm surges that can cause severe flooding.

Weather Words *(Cont'd.)*

Lightning   The visible discharge of electricity associated with a thunderstorm.

Low pressure   An air mass associated with storms.

Meteorology   The science that studies the weather.

Mostly cloudy   Condition when 70% to 90% of the sky is covered with clouds.

Mostly sunny   Condition when 10% to 30% of the sky is covered with clouds.

Overcast   Condition when 90% or more of the sky is covered with clouds.

Partly cloudy (or **partly sunny**)   Condition when 30% to 70% of the sky is covered with clouds.

Precipitation   Any form of water, including rain, drizzle, sleet, hail, or snow, that falls to the ground.

Rain   Precipitation consisting of droplets of water.

Sleet   Precipitation consisting of ice particles.

Smog   Fog mixed with particles of pollution.

Snow   Precipitation consisting of small ice crystals.

Sunny (or **clear**)   Condition when less than 10% of the sky is covered with clouds.

Temperature   Degree of hotness or coldness of the air.

Thunder   The sound caused by lightning heating air and causing it to rapidly expand.

Tornado   A violent whirlwind often accompanied by heavy rain, hail, thunder, and lightning.

Variably cloudy   Condition when cloudiness of the sky varies between 20% and 90%.

Weather   The condition of the atmosphere at a given time.

Wind   Moving air.

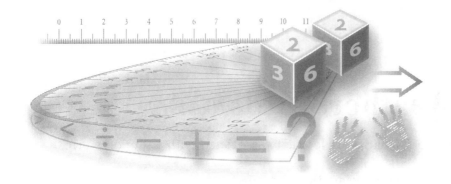

A flight to Mars

A human expedition to Mars has been the dream of space enthusiasts for years. Probably within the next twenty-five years, a human flight to Mars will be launched. As the baby boomers witnessed the moon landings, their children (and grandchildren) will see astronauts walking on Mars. Even today, long before that flight becomes a reality, we can speculate on what it will be like. Your students can do some speculation for this project too.

Goal

Working in groups of three to five, students will assume that their group is part of the first human flight to Mars. They will further assume that the entire expedition will last about two and a half years. As crew members, their group shares a cubicle that is 2 feet by 2 feet by 2 feet in which they may store personal items on the trip. They are to decide which items or equipment they would bring, being sure that their materials will fit within the cubicle. At the end of the project, each group will need to share their list and explain why they selected the materials they did. *Suggested time:* Two to three class periods.

Math Skills to Highlight

1. Making a model
2. Visualizing spatial relationships
3. Using measurement to determine dimensions and volume

Special Materials/Equipment

Rulers, tape, cardboard, tape measures, poster paper, graph paper, thin markers, crayons, colored pencils. *Optional:* Reference books about Mars; computers with Internet access for research.

Development

Tell your students that within their lifetimes, it is probable that humans will journey to Mars in exploration. The trip will be long and dangerous, the conditions on board will be cramped and mostly uncomfortable, and there will not be any fast-food outlets along the way. Given those conditions, ask your students why they think people would go. You might mention that the conditions were probably tougher for the early comers to the New World.

- Start this project by explaining that each group is to assume that it will be part of the crew of the first human expedition to Mars. Because space on board will be very limited, they will share a cubicle, 2 feet by 2 feet by 2 feet, into which they may store personal belongings. Materials for basic needs such as food, water, clothing, toothpaste and brushes, and exercise equipment are supplied for them. Personal items might include books, small CD and DVD players, CDs and DVDs, batteries, games, and maybe even musical instruments.

- Hand out copies of Student Guide 10.1, and review the information with your students. Make sure that they understand that they do not need to bring along items or equipment that satisfy basic needs of food, shelter, clothing, exercise, or personal hygiene.

- Using poster paper, cardboard, or similar materials, construct a model of a 2-foot cube so that your students can visualize the size of the space in which they will be able to store items. Use tape to secure the sides. You might have a group of students do this. (You can also use a cardboard box with similar dimensions.)

- Hand out rulers to students for measuring items. Encourage them to measure items at home, because it is unlikely they will have the items in class they would like to take on such a flight. Tape measures are also useful. (Many students probably have access to rulers and tape measures at home.)

- Emphasize that group members are to share the cubicle. They should agree as to what items or equipment to bring. Packing is an important consideration.

- Point out that the items they select should be practical. Items that provide benefits to several members of the group or crew should be given priority.

- Distribute copies of Data Sheet 10.2, "Information About the Red Planet," which contains facts about Mars that will help students understand Mars's environment. Suggest that students consult other references for more information.

- After selecting the items they would bring, your groups are to draw what they think their cubicle would look like once it is packed. They should use graph paper, which will make it easier for them to create and use a scale. Suggest that they draw the cubicle from a front perspective, although they might like to draw other perspectives as well. They may illustrate their work with markers or colored pencils. They should list their items on the side of their illustration. Depending on the type of drawing, they may not be able to show all of the items inside their cubicle.

Wrap-Up

Each group will display their poster, and the group's spokesperson will explain why the group selected the items they did.

Extension

Suggest that students consult additional sources on Mars to find out what a flight to the planet might entail. Have them report back to the class with their findings.

STUDENT GUIDE 10.1

A Flight to Mars

Situation/Problem

Your group will be among the crew of the first human expedition to the planet Mars. Leaving Earth when Mars and the Earth are making their closest approach to each other, the flight will last about six months. The crew will have to remain on Mars for nineteen months until Mars and Earth approach each other again; they will spend another six months on the flight returning home. The entire expedition will take about two and a half years. All of your group's basic needs—food, water, clothing, items for personal hygiene, and exercise equipment—will be provided. Because the space on board is limited, each group will be assigned a cubicle 2 feet high, 2 feet wide, and 2 feet deep in which members may store personal items. Your team must work together to agree on a list of items you would take to Mars. The items must be able to fit within the dimensions of the cubicle.

Possible Strategies

1. Brainstorm a list of items and equipment. Then go back over the list and eliminate items in order to fit everything in the space available. Remember that your group must share the space in the cubicle.

2. Consider bringing items that everyone can use. For example, books may be read by everyone.

3. Try to avoid items that will take up a lot of area unless they are extremely important to the group.

A Flight to Mars *(Cont'd.)*

Special Considerations

- Obtain the measurements of items you would like to bring. Divide the task of measuring among group members. If each of you measures five items, the overall task of measuring will not be as time-consuming.

- Although dimensions are important, weight is not. Your ship will be built in space where there is no gravity. The gravity on Mars is 38% of Earth's.

- Remember that your trip will last about two and a half years. Select items that will last.

- Refer to Data Sheet 10.2 for information about Mars. Consult other references for more information, including online sources.

- Sketch or draw your cubicle, and then sketch the items inside. Work with a scale, something like 2 inches equal 1 foot. This will help you to visualize how items would fit in your cubicle. Make any final changes.

- After you have selected the items your group would take, draw a final example of your packed cubicle. Use graph paper, and create a scale. Several lines on the paper might equal 1 foot. Depending on the perspective you use—whether you draw your cubicle from the front or side, for example—you may not be able to show all of the items inside. Color your picture, and list the items along the side.

- Pick a spokesperson to share your poster, and explain why you chose the items you did.

To Be Submitted

Picture with list of selected items.

Information About the Red Planet

Mean distance from sun: 141.3 million miles (227.9 million kilometers).

Length of year (in Earth time): 687 days.

Length of day (in Earth time): 24 hours, 37 minutes.

Diameter: 4,200 miles (6,800 kilometers).

Number of moons: Two (Phobos and Deimos).

Mean surface temperature: –9 degrees F (–23 degrees C).

Surface gravity (compared to Earth): 0.38

Main gases in atmosphere: Carbon dioxide, with small amounts of nitrogen, oxygen, and argon; however, the Martian atmosphere is very thin.

Landscape: Reddish rock and dust, a result of iron oxide. This is where the name "Red Planet" comes from. The land is dry, dotted with towering volcanoes, and criss-crossed with "canals." Once thought to be real canals dug by Martians, most planetary scientists now believe the canals are dry riverbeds and streams. Most agree that Mars once had flowing water.

Potential habitat for humans: Living on Mars would mean existing in temperatures much like those found on Antarctica. Humans would also need pressurized suits and oxygen. The first human bases on Mars would probably be constructed underground.

Why go to Mars? Mars is probably the most hospitable planet in our solar system other than Earth. Many scientists believe that water is trapped beneath its surface. If this is true, that water can be used by future astronauts. Some scientists believe that varieties of Arctic plants may be able to grow on Mars. This could be a source of oxygen. Along with revealing information about the formation of our solar system, Mars could one day be a colony for humans.

An interesting note: Any humans being one day born on Mars would officially be Martians.

Math and Social Studies

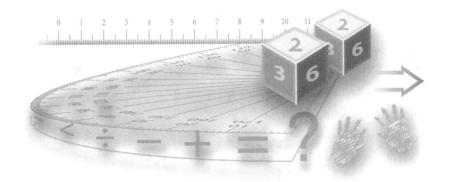

A great mathematician

Throughout history, numerous men and women have made significant contributions to the study and advancement of mathematics. Unfortunately, the contributions of many are largely forgotten because society tends to view math in terms of its applications and not its history. This project gives your students the chance to learn about some of the individuals who are responsible for our use and understanding of mathematics today.

Goal

Working individually, students will research and write a report about a mathematician of their choice. After writing their report, they will make an oral presentation to the class offering the major accomplishments of the mathematician they selected. *Suggested time:* Two to three class periods over two weeks.

Math Skills to Highlight

1. Researching information about a mathematician
2. Evaluating information and drawing conclusions
3. Organizing information
4. Using mathematics to communicate ideas
5. Using technology in problem solving

Special Materials/Equipment

Reference books; note cards. *Optional:* Computers with Internet access for research and writing reports; PowerPoint or an interactive whiteboard for presentations.

Development

Discuss with your students that mathematics has been advancing for thousands of years, developing to meet the needs of people and society. One of the earliest uses of math was likely counting on fingers during bartering. There are records indicating that by 3000 B.C., ancient Egyptians and Babylonians had developed number systems and used early forms of bookkeeping in trade. Many people and groups have contributed to the development of math since then.

- Begin this project by explaining that students will work alone and research a mathematician of their choice.
- Note that students are to do a written report and provide an oral presentation about their mathematician.
- Suggest that they research their subject's background, education, and specific contributions to the advancement of mathematics. If possible, they should include examples, such as equations, figures, or specific ideas, that demonstrate some of his or her work.
- Distribute copies of Student Guide 11.1, and review the information with the class. Suggest that students consult both print and online sources for their research. It is possible that the mathematician they are researching will have some Web sites that include his or her achievements.
- Warn students that it will be easier to find information on some mathematicians than others. They should be persistent.
- Hand out copies of Data Sheet 11.2, "Math Greats." Review the men and women listed on the sheet, and suggest that students choose one of them for research. You may wish to expand this list and allow students to

116

choose other mathematicians, but instruct them to obtain your approval before they begin researching.

- To help students with their research, you may take them to the school library for a period or two. Inform the librarian in advance that the class will need books on mathematicians. Suggest that students also visit the public library and check online sources.

- Note that much of the research and work on their reports will be done outside class.

- After concluding their research and writing their reports, students are to provide an oral presentation to the class that highlights the major points of their reports and, if possible, offers examples of the mathematician's work. Consider having students incorporate the use of an overhead projector, PowerPoint, or an interactive whiteboard in their presentations.

Wrap-Up

Students make their presentations.

Extension

Organize groups and instruct students to create a time line highlighting significant achievements in mathematics.

A Great Mathematician

Albert Einstein

Situation/Problem

You are to select a mathematician and research his or her life and contributions to mathematics. When you are done, you are to write a report summarizing your research and provide an oral presentation to the class.

Possible Strategies

1. Select a mathematician who has done work on a topic in which you are interested.
2. Consult reference books, biographies, and online sources to find information about your subject.

Special Considerations

- Focus your research efforts on your subject's life, background and education, and contributions to math. Ask yourself how those contributions benefited other mathematicians or people in general.

- Take accurate notes, and record your sources on your notes.

- Include examples that demonstrate or highlight some of your mathematician's contributions. Consider equations, illustrations, figures, or examples of problems.

- When writing your report, be sure to include an introduction, a body with main ideas and supporting details, and a conclusion.

- Offer your opinion. Was this person a great mathematician? Or was he or she overrated? Explain your answer.

- Be sure to include complete bibliographical information for both print and online sources. Consult your English text or an author's stylebook for the correct format.

A Great Mathematician *(Cont'd.)*

- When planning your oral presentation, prepare an outline and provide background information on your mathematician, but focus most of your presentation on his or her contributions to math. If possible, offer examples.

To Be Submitted

1. Your report
2. Your notes
3. The outline of your presentation

Rough Notes

Math Greats

Following are some of the men and women who have made significant contributions to the advancement of mathematics. There are many more. The list notes the nationality and areas of major accomplishments of these men and women.

Abel, Niels (1802–1829), Norwegian; algebra

Ahmes (about 1650 B.C.), Egyptian; geometry

Aiken, Howard (1900–1973), American; computers

Al-Khowârizmî, Muhammed (about 780–850), Arabian; algebra

Archimedes (287–212 B.C.), Greek; algebra, calculus, pi

Aristotle (384–322 B.C.), Greek; logic, geometry

Celsius, Anders (1701–1744), Swedish; measurement

Copernicus, Nicolaus (1473–1543), Polish; trigonometry

Cray, Seymour (1925–1996), American; computers

Descartes, René (1596–1650), French; coordinates

Dodgson, Charles L. (Lewis Carroll, 1832–1898), English; logic

Einstein, Albert (1879–1955), German; geometry, infinity

Escher, Maurits Cornelis (1898–1971), Dutch; geometry

Euclid (about 365–300 B.C.), Greek; geometry

Fahrenheit, Gabriel (1686–1736), German; measurement

Fermat, Pierre de (1601–1665), French; number theory

Gauss, Carl Friedrich (1777–1855), German; geometry, number theory

Germain, Sophie (1776–1831), French; symmetry

Hypatia (370–415), Greek; conic sections

Kovalevsky, Sonya (1850–1891), Russian; number theory

Leibniz, Gottfried (1646–1716), German; logic, calculus

Leonardo da Vinci (1452–1519), Italian; geometry

Murasaki, Lady (about 978–1031), Japanese; combinations

Napier, John (1550–1617), Scottish; computers, decimals

Newton, Sir Isaac (1642–1727), English; algebra, calculus

Noether, Emmy (1882–1935), German; algebra

Oresme, Nicole (1323–1382), French; functions

Pascal, Blaise (1623–1662), French; algebra, computers

Ptolemy, Claudius (about 85–168), Greek; trigonometry

Pythagoras (about 585–507 B.C.), Greek; geometry

Romanujan, Srinivasa (1887–1920), Hindu; algebra

Venn, John (1834–1923), English; sets

Von Neumann, John (1903–1957), Hungarian; computers

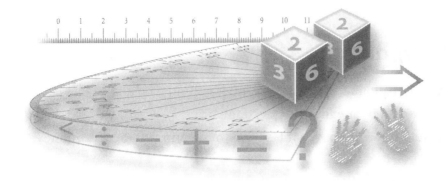

An election poll

Polls, especially around election time, are a part of our lives. Polls show people's opinions about issues, how they feel about policies, and how they might vote. At their simplest, a poll might be little more than a few questions that a researcher asks randomly selected individuals. Sophisticated polls consist of carefully designed questions targeted at specific groups. This project gives your students a chance to conduct an election poll.

Goal

Working in groups of four or five, students will write at least five questions on political issues, concerns, and candidates in an upcoming election. The class as a whole will select the best questions for inclusion on a questionnaire that they will distribute to potential respondents. After the questionnaires have been collected, students will work in their groups to construct frequency tables and analyze the results. A class composite result will be expressed as a percentage. *Suggested time:* Two to three class periods over two weeks.

Math Skills to Highlight

1. Writing a questionnaire, the results of which may be analyzed
2. Gathering data by means of a questionnaire
3. Constructing frequency tables and determining the frequency of responses
4. Using percentages to show information
5. Making a prediction based on collected data
6. Using technology in problem solving

Special Materials/Equipment

Calculators for tabulating results and finding percentages. *Optional*: Computer and printer for writing and printing out the questionnaires; a clipboard to use for conducting the survey; Internet access for research.

Development

Becoming involved with political polls is an excellent way for students to learn about issues that may influence the outcome of elections. The best time to assign this project is a few weeks before the November elections. You may focus the project on presidential, gubernatorial, state, or local elections, or a combination of the different races and the issues that accompany them. Since it is likely that students will be following the elections in their social studies class, you might wish to share this project with your students' social studies teacher. She might incorporate it into her plans, expanding the project and its benefits.

- Start the project by explaining to students that they will work in groups to develop questions for an election poll. Each group will contribute questions to a questionnaire.

- Note that there are many ways to conduct polls. They may be conducted over the telephone, through the mail, or with a personal interview. Sometimes the questions in a poll are designed to be answered with a simple "yes," "no," or "undecided"; sometimes people are asked for explanations; and sometimes they are asked to rate, or prioritize, issues or concerns.

- If you are working with your students' social studies teacher, he might be able to discuss some of the issues of the upcoming election with your students before they begin this project. If you are working alone, spend some class time discussing the issues, and encourage students to read about the campaigns on their own.

- Hand out copies of Student Guide 12.1, and review the information with your students. Make certain they understand that each group is to contribute to the class's overall efforts. (Each group will write questions for the poll; then the class as a whole will pick the best ones.) Especially note that students should consider using questions that may be answered with a "yes," "no," or "undecided." These are easiest to interpret. While questions may ask respondents to rank issues, such questions are harder to interpret. Depending on your class, you may instruct students to avoid such questions.

- Distribute copies of Data Sheet 12.2, "Poll Points," and go over the material with your students. The sheet offers information about polls, as well as some sample questions and suggestions on how to conduct the poll.

- We recommend that you review the drafts of each group's questions before the group members write their final copies. This enables you to offer suggestions if necessary.

- After each group has written its questions, work together as a class to select the five to ten questions that will be on your classroom questionnaire. Try to include at least one question from each group. A good way to select the questions is to write them on an overhead projector and discuss the merits of each. Some questions may ask similar information. In this case, discuss what makes one question better than the other.

- Once the questions have been selected, ask for a volunteer to produce a copy of the questionnaire. Using a computer makes revision easier. (Polls often ask respondents information such as their age, sex, and political affiliations. If you wish to expand the scope of your questionnaire, ask the class if they wish to collect such data. If they do, include such questions on the questionnaire.)

- Copy enough questionnaires so that each student has ten to distribute. The more respondents the class obtains, the more accurate the results of their poll will be. If your class has twenty-five students and each one hands out ten questionnaires, you should get enough back to have a large sample.

- Explain to students that they might ask friends and relatives to answer their questionnaire. They should not go door to door or ask strangers.

- Set a deadline for the completed questionnaires to be returned. Five to seven days is reasonable.

- After the questionnaires have been returned, students should work in their groups to tabulate the results of their questionnaires. (Each group works with the questionnaires its members distributed.) Suggest to students that they create a frequency table to record the results for each question.

- After each group has analyzed the results for its questionnaires, you should compile the results for the class. You might do this on an overhead projector. Have the class help you express the overall results as percentages; for example, "65% of respondents answered 'yes' to question 1." If the class gathered data about the respondents, you may wish to analyze the answers according to these different categories.

Wrap-Up

Discuss the results of the poll with your class. Based on the results, make predictions for the upcoming election.

Extension

Suggest that students continue to follow the election and pay close attention to how polls play a role in campaigns. Polls may even influence the outcome of the election.

STUDENT GUIDE 12.1

An Election Poll

Situation/Problem

Your group is to develop three to five questions about the issues or candidates during an upcoming election. After writing your questions, you will work with the rest of your class to select five to ten questions that will be included in a class questionnaire that students will distribute. After gathering data, each group will take part in analyzing the results of the poll.

Possible Strategies

1. Learn about the upcoming election by reading newspaper or magazine articles and watching news broadcasts. You may also check online sources.

2. Discuss campaign issues with your group.

3. Brainstorm possible survey questions with your group. See Data Sheet 12.2 for examples.

4. Choose what your group feels are your best questions.

5. Keep your questions simple. If they ask for too much information, you will have a hard time analyzing the answers.

An Election Poll *(Cont'd.)*

Special Considerations

- After each group has written questions, you will meet with the entire class and select five to ten questions that the class agrees are the best. These questions will be compiled in the form of a questionnaire. Your teacher will provide class members with several copies of the questionnaire.

- Each student should distribute the questionnaire to ten people. Survey only friends and relatives. Do not go door-to-door or survey strangers. When you hand out the questionnaire to people, ask them to complete it while you wait. If they want to do it later, they may forget or not have the time. Do not pressure them, though. If they need more time, pick the questionnaire up later. Handing your questionnaire on a clipboard to people encourages them to answer the questions immediately. Have a pen or pencil handy.

- After obtaining your results, your group should make a frequency table like the one shown here to record them. Here is how to make a frequency table for questions that require a "yes," "no," or "undecided" response:

  Make a table with three columns.

  List the possible answers ("yes," "no," "undecided") in the first column.

  Make a tally mark in the second column.

  Total the tally marks in the last column.

Response	Tally	Frequency
Yes		
No		
Undecided		

- A frequency table for questions that asks respondents to rank issues may be done in a similar manner. For each ranking, a separate tally would be marked.

- After each group has analyzed its results, the entire class will compile the results and use percentages to express the data.

To Be Submitted

1. A copy of your group's questions
2. A frequency table showing the results for each question on the questionnaire

DATA SHEET 12.2

Poll Points

Following are some tips that will help you write and conduct an effective election poll:

- The more respondents you have, the better your chances are for accurate results. A poll that has 500 respondents will show more accurate results than a poll of 20.

- Carefully consider the questions you will ask for your poll. Questions that can be answered with a "yes," "no," or "undecided" are easiest to interpret. Questions that ask respondents to rank issues or concerns in the order of importance provide more information but are usually harder to analyze. Think about how to phrase the questions and the possible ways to construct the frequency table. Consider the use of "other" and "none" as responses.

  An example of a "yes" or "no" question: "In the upcoming election, will you vote yes or no for the referendum on building a new school in town?"

  Yes _____   No _____   Undecided _____

  An example of a question that asks respondents to rank issues: "Rank the following in importance to you (1 being of most importance and 5 being of least)."

  Crime _____   Job Security _____   Health Care _____   Taxes _____   Other _____

- Many polls ask respondents to note their age, sex, or political affiliation. This information helps pollsters interpret the results according to specific groups. You may want to include such information on your poll.

- Questionnaires should be printed clearly and should be easy to read. Have a clipboard and pen handy to make it easier for people to answer your questionnaire.

- When you distribute questionnaires to potential respondents, be polite. If someone is not interested, do not press; simply thank him or her and move on.

- When someone is filling out your questionnaire, give her time. If a person does not have the time now, ask her to return the questionnaire to you by a specific day. Since you will have distributed questionnaires to people you know, you should plan to go to them to collect the questionnaires. This will speed results. Set a deadline a few days ahead of when you really need the questionnaire. Remember that everyone's time is valuable.

- After people answer your questionnaire, thank them. After all, they are doing this for you.

Poll Points *(Cont'd.)*

Some Useful Vocabulary When Conducting a Poll

Data   Facts and figures that we can use to obtain information.

Frequency   The number of times an event occurs.

Frequency table   A table that organizes the results of a tally so that the frequency of each can be recorded.

Poll   A survey, often one asking people how they will vote in an election.

Pollster   An individual conducting a poll.

Sample   A small group used to provide information about a larger group. For example, it would be impractical (if not impossible) to ask every person in a particular state which candidate they will likely vote for in an upcoming election. By asking 1,000 random voters, however, pollsters can make a prediction how the election in the state will turn out.

Statistics   Data collected and arranged in a systematic manner.

Survey   A method of collecting data from a sample.

Tally   A count or total of specific responses to a question.

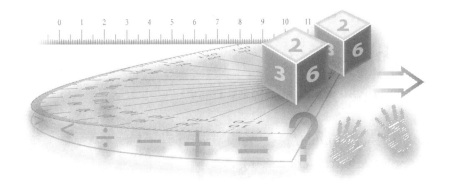

Landmarks

Depending on the curriculum and grade level, various countries, or, in the case of the United States, different states are studied in social studies. In this project, students will find and enlarge pictures of landmarks of a specific country or state. The enlarged pictures may be used for bulletin boards, hallway displays, as part of multicultural activities, or as pictorial displays of countries or states.

Goal

Working individually, students will consult reference books and online sources to find a picture of a landmark of a specific state or country. They will enlarge the landmark proportionally. *Suggested time:* Two class periods.

Math Skills to Highlight

1. Measuring and dividing line segments
2. Locating points on a grid
3. Identifying lines of symmetry
4. Using technology in problem solving

Special Materials/Equipment

Encyclopedias, social studies textbooks, and other reference books; tracing paper; rulers; colored pencils; markers; white drawing paper. *Optional:* French curve; computers with Internet access for research.

Development

You may wish to work with your students' social studies teacher on this project. If your school has a multicultural committee, its members might be interested in coordinating this project with other activities. Before introducing the project to your students, consult their social studies teacher for countries or states that they will be studying throughout the year. You may wish to compile a list from which students may choose, or you may open the project up a bit and let them select any country or state that interests them. Many students will choose countries of their ethnic origins. It is not necessary for every student to choose a different country.

- Begin this project by explaining to your students that they will select a specific country or state, obtain a picture of one of its major landmarks, and then enlarge the landmark in an exact proportion so that the finished enlargements will be 6 by 9 inches. Examples of landmarks include the Statue of Liberty and Washington Monument in the United States, Big Ben (United Kingdom), the Eiffel Tower (France), the Leaning Tower of Pisa (Italy), and the Taj Mahal (India).
- Encourage students to select pictures of landmarks that are relatively simple and free of elaborate details. This will make it easier to complete the enlargement.
- Distribute copies of Student Guide 13.1. Review the information with your students.
- Distribute copies of Data Sheet 13.2, "Steps to Enlarging a Pattern." Carefully review with your students the instructions for increasing the size of the pictures of their landmarks. Depending on your class, you may feel that you should demonstrate the process.

- Schedule a period in the library for students to find landmarks they would like to enlarge. You might also have students search online for landmarks. In this case, they will need to download the picture of the landmark and print it. As students are researching, hand out rulers, tracing paper, and colored pencils.

- Once students have selected their landmark, they should trace it. Remind them to be as accurate as possible during tracing. (If you have access to a photocopier, consider copying their landmarks.) If students conduct their research in the library and the rest of the project is to be completed in class, instruct them to lightly color (or label the colors of) their traced picture. Emphasize the need to color lightly because they will have to draw lines over the picture later.

- Discuss lines of symmetry with your students. Note that symmetrical lines make enlargements easier because one side of a symmetrical figure is a mirror image of the other side.

- Encourage students to use rulers for straight lines and a french curve for irregular curves. Some students, especially good artists, may wish to sketch the intricate details of their landmarks by hand.

- After enlarging their landmarks, students should color them according to the landmark's actual colors.

Wrap-Up

Display the finished landmarks throughout the school, or in math or social studies rooms.

Extensions

Students may wish to enlarge other items such as maps or flags or create their own patterns for enlargement.

STUDENT GUIDE 13.1

Landmarks

Taj Mahal

Situation/Problem

Landmarks can be powerful symbols of a people, their heritage, or their beliefs. Think of the Statue of Liberty and all that it means. In this project, you will select a country or state, obtain a picture of a major landmark, and enlarge and color it.

Possible Strategies

1. Choose a country or state in which you are interested, and obtain a picture of a distinctive landmark. This might be a famous building, statue, bridge, or equally significant structure.

2. Look for lines of symmetry to simplify your drawing. This will help you to make an enlargement that is proportional to the original picture by enlarging only part of the landmark. You may use the enlarged portion as a guide to sketch the other part of the drawing, since one part is symmetrical to the other.

Special Considerations

- Follow all directions on Data Sheet 13.2 carefully.
- Choose a picture that is relatively free of details or whose outline is relatively simple. Too many details may make the landmark difficult to enlarge:

  Make sure that the lines of your landmark are straight. Detailed lines will be hard to enlarge accurately.

  Be sure your vertical and horizontal lines are straight.

Landmarks *(Cont'd.)*

Check the numbers you will use to enlarge your landmark. Are they arranged chronologically?

Check the letters. Are they in alphabetical order?

Do your numbers and letters correspond to lines and not spaces?

Plot the points carefully. Estimate if necessary.

Draw or sketch irregular lines.

- If the landmark you chose has any lines of symmetry, use the lines of symmetry to guide you as you enlarge the pattern.
- Use colored pencils or markers to color your landmark.

To Be Submitted

1. Original picture or tracing
2. Enlarged drawing of your landmark

Notes

_____

_____

_____

_____

_____

_____

_____

_____

_____

_____

_____

_____

Steps to Enlarging a Pattern

1. Trace the picture of the landmark you wish to enlarge from a reference book, or download it from the Internet. Lightly color or color-code the picture of the landmark. You can color-code it by using a letter; for example, R for red, B for black, and Br for brown. This saves you the time needed for coloring, although coloring will give a more accurate representation.

2. Obtain a plain piece of $8\frac{1}{2}$- by 11-inch white paper on which you will draw your enlarged landmark. Draw a 6- by 9-inch rectangle horizontally on the page. This will be the full-sized pattern of your enlarged landmark.

3. Divide the full-sized pattern into 1-inch squares. Draw these lines *lightly* in pencil. Be accurate.

4. Start at the lower left-hand corner of your full-sized pattern, and number 1 to 10 across the bottom. Place each number directly below a line, not between lines.

5. Start at the lower left-hand corner of your full-sized pattern and letter from A to G, moving upward. Each letter should be placed directly before a line, not between lines.

6. Now take the tracing of your landmark. Divide the tracing into 9 equal sections, moving from left to right across the bottom, and numbering from 1 to 10 in the same way you labeled the full-sized white paper. Draw the vertical lines.

7. Divide the picture of the small landmark into 6 sections vertically, and letter them from A to G, as in the full-sized pattern. Draw the horizontal lines.

8. Locate points on the small landmark, and transfer them to the large grid. Continue to do this by including all points in the intersections. Estimate the placement of several other points until a shape is outlined.

9. Fill in the other points by sketching or using a french curve. To use a french curve, do the following:

    Locate 3 points on a section of the curve. This may require some trial and error.

    Draw the curved section up to, but do not include, the third point.

    Move the curve, and repeat this procedure until all irregular curves are sketched.

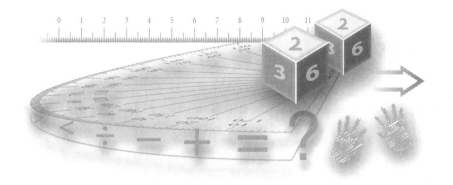

Creating a scale map

The ability to read maps is a skill that most people use throughout their lives. While students have many opportunities to copy and read maps, few ever create them. When they do, they realize that mathematics plays a major part in mapmaking.

Goal

Working in pairs or groups of three, students will create a scale map of their school, school grounds, or yard at the home of one of the group members. They will include landmarks, important details, legends, and an accurate scale. *Suggested time:* Two class periods.

Math Skills to Highlight

1. Measuring and rounding distances
2. Determining an appropriate scale
3. Making a scale drawing

Special Materials/Equipment

Measuring tapes, yardsticks, meter sticks, or trundle wheels; compasses (for finding directions); rulers; colored pencils; markers; poster paper (approximately 22 by 28 inches).

Development

We suggest that you give students options regarding the maps the groups would like to draw. Students may wish to measure and create a map of the yard of one of the members of their group; students who live in apartments might prefer to measure and create a map of their school or school grounds. If students are to draw a map of their school, you might decide to limit the map to a section or wing.

- Begin this project by explaining that maps use a scale to ensure accurate distances. On a world map, 1 inch might equal 1,500 miles, while on the map of a small town, 1 inch might equal $\frac{1}{4}$ of a mile. On a map of a yard or block, 1 inch might only equal several feet. Scales are created with mathematics.

- Distribute copies of Student Guide 14.1, and review the information with your students.

- If you wish, you may simplify the project by instructing students to create a scale map of your school or school grounds. In this way, you may take your class to obtain the necessary measurements together.

- In measuring distances, remind students to be as accurate as possible. We suggest that distances be rounded to the nearest foot, nearest yard, or nearest meter.

- Suggest that students sketch a rough copy of their map first. This will help them to see relationships and estimate where things should be.

- Distribute copies of Data Sheet 14.2, "How to Make a Map." Go over the material with your students, making sure that they understand how to create a scale for their maps. You may want to check the scales before the students start making the final copies of their maps.

- After students draw their maps, they should label and color them. Suggest that they provide legends, label the directions, and note the scale.

Wrap-Up

Display the maps of your students.

Extension

Encourage students to pursue a study of an instrument called the architect's scale, and use it to make a scale drawing of their house. If their house has two stories, they may need to create more than one drawing.

STUDENT GUIDE 14.1

Creating a Scale Map

Situation/Problem

You and your partner or group are to create a scale map of a familiar place such as your school, school grounds, or the yard of your home or the home of your partner or another group member.

Possible Strategies

1. Accurately measure distances (rounding to the nearest foot, yard, or meter).

2. Note landmarks. In the case of a yard, this might include trees, woodpiles, a shed, or a swing set. You might include these in a legend on your map.

3. Create a rough sketch of your map before drawing a final copy. A "rough" will help you to visualize perspectives and landmarks.

Special Considerations

• Use a measuring tape, yardstick, meter stick, or trundle wheel for measuring distances.

• Use a pad and pencil to record distances. Do not try to remember them; you will likely make mistakes if you do.

• As you record distances, sketch your map, placing landmarks about where they would be. Record the distances in feet, yards, or meters. It is a good idea to locate landmarks using the measurements from two boundaries.

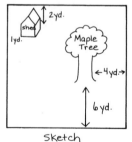

Sketch

• Use a compass to find directions. Be sure to label the directions correctly on your map.

• Consult Data Sheet 14.2 for information about working with scale drawings.

138

Creating a Scale Map *(Cont'd.)*

- Be sure the final copy of your map is accurate. Label distances and landmarks, add color, and include a legend and directions. You may want to compare your map to the original area and check it for accuracy.

To Be Submitted

1. Your map
2. Your records of measurements and calculations

Rough Sketch

How to Make a Map

1. Decide on the boundaries of your map.

2. Make a sketch of the area that you will include on your map. Note the approximate position of any landmarks. In a school, landmarks might include stairwells, display cases, or water fountains. Landmarks in a yard might include trees, flower beds, decks, sheds, woodpiles, or swing sets.

3. Accurately measure the boundaries (length and width) of the area. Locate the position of landmarks by obtaining at least two measurements from boundaries.

4. Select the scale by considering your longest measurement and how to fit it on the paper. Remember that the scale should be as long as possible so that your map will look good on the paper.

5. To choose the best scale, divide the longest length of your paper in inches (or centimeters) by the longest dimension of the boundary in feet (or meters).

   Round your quotient *down* to the nearest quarter- or eighth-inch (or centimeter). (If you round up, your map might not fit on your paper.) Here is an example: The longest boundary on your map is 80 feet. The longest dimension of your paper is 28 inches: $\frac{28}{80} = 0.35$. Since 0.35 is between 0.25 ($\frac{1}{4}$ inch) and 0.375 ($\frac{1}{8}$ inch), you must round down so that your scale will be $\frac{1}{4}$ inch = 1 foot.

   Now take the other dimension of the boundary and the other dimension of the paper, and divide the length of the paper by the length of the boundary.

   Round your quotient down to the nearest quarter- or eighth-inch (or centimeter).

   Compare the scales. If they are the same, great! If they are different, use the smaller scale.

6. To place items on your map, use your measurements and the scale you have chosen. For example, suppose an apple tree is 21 feet from the fence on the eastern side of the yard, and 16 feet from the fence on the northern side. If your scale is $\frac{1}{4}$ inch = 1 foot, multiply the number of inches by the number of feet to determine the number of inches the actual distance would be on your map. Note the example below:

$$\frac{1}{4} \times \frac{21}{1} = \frac{21}{4} = 5\frac{1}{4}$$
$$\frac{1}{4} \times \frac{16}{1} = 4$$

On your map, place the tree $5\frac{1}{4}$ inches from the fence on the eastern side and 4 inches from the northern side.

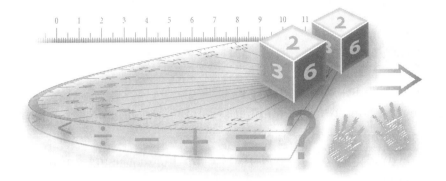

A game: What date is that?

When your students are learning about the ancient civilizations of the Egyptians, Greeks, and Romans in social studies, you may wish to incorporate a study of the number systems of these civilizations in your math class. Learning about ancient number systems can help students gain an understanding of the evolution of mathematics and appreciate the Hindu-Arabic system we use today.

Goal

After learning about ancient number systems, students, working in pairs or groups of three, will use these number systems to create a time line of important dates in the history of the civilization they select. The time lines will contain only dates, not the events that go with them. The time lines can then be photocopied and distributed to the class, whose task will be to convert the dates to the Hindu-Arabic number system we currently use and match the dates with their events. *Suggested time:* Two class periods.

Math Skills to Highlight

1. Writing numbers using other number systems
2. Comparing the Hindu-Arabic number system to ancient number systems
3. Determining units of a time line
4. Creating a time line of important dates in history using an ancient number system
5. Using technology in problem solving

Special Materials/Equipment

History books and other reference books containing information about the ancient civilizations; black felt-tipped pens; rulers; $8\frac{1}{2}$- by 11-inch white paper for drawing the time lines. *Optional:* Computers with Internet access for research.

Development

Check with your students' social studies teacher in advance to find out when she will begin teaching ancient civilizations. Suggest to her that you would like to teach your students about the number systems of these civilizations. Of course, you may assign this project even if your students are not studying ancient civilizations. If that is the case, we suggest that you encourage your students to consult books in the library about the ancient Egyptians, Greeks, and Romans, particularly their number systems. They may also consult online sources.

- Begin this project by explaining to your students that many ancient civilizations have contributed to the development of mathematics and the number system we use today.
- Note that we use the Hindu-Arabic number system, which was developed over twelve hundred years ago in the Middle East and India. This number system was not introduced to Europe until the 1100s.
- You may discuss any or all of the three number systems included in this project: the ancient Egyptian, Greek, and Roman. If you wish to limit the project, select one or two. However, using all three will result in a greater variety of time lines, which will make the wrap-up activity more interesting. (Since these ancient civilizations are often taught in succession in many history classes, a good time to assign this project might be as soon as all of the civilizations have been studied.)

- Distribute copies of Student Guide 15.1, and review the information with your students.
- Distribute copies of Data Sheet 15.2, "Ancient Number Systems," which shows how people in the selected civilizations wrote numbers. Review the procedure for working with numbers for each civilization as outlined on the sheet. You might want to write a few dates on an overhead projector or the board and have students convert them to the number systems of the ancient civilizations. This will be good practice.
- Explain that students should designate B.C. where necessary for dates. Although ancient civilizations did not use B.C. with their calendars, since students will be converting the dates of the ancient civilizations to our modern Hindu-Arabic number system, the B.C. notation is necessary.
- Instruct students to pick one of the number systems, list at least five important dates in the history of that civilization, and then express those dates using the ancient number system. If you wish to expand the project, suggest to students that they may do more than one number system. (Prior to this class, ask students to bring their history books to your math class. They will probably need a reference to identify accurate dates. You may also consider obtaining additional references from the school library, as well as suggest that students search online sources.)
- Instruct students to create a time line using the dates they selected. They are to write the dates in the number system they chose. Remind them not to list the event next to the date but instead leave a blank line. Their classmates will interpret the date and write the event on the line. Depending on your class, you may want to review how to construct a time line so that the dates will correspond with points on the number line.

Wrap-Up

Photocopy the time lines, and distribute them to the class. Students can then try to work out the dates and match them with the correct events. This can be an enjoyable activity.

Extension

Encourage students to select a fictional civilization (from literature or their own imaginations), and create a number system for this civilization.

A Game: What Date Is That?

Situation/Problem

Imagine that you and your partner or group are students in ancient Egypt, Greece, or Rome. Like students today, you have assignments to do and facts to learn. You are to select *at least five* important dates in the history of your civilization, and create a time line. You must write the dates using the number system of your civilization. Do not label the events on your time line. Simply leave a line next to the date for the event, and let your friends convert the dates to our number system. They should then try to match events to the dates.

Possible Strategies

1. Discuss with your partner or group which ancient civilization you would like to work on for this project.

2. Use a reference book or your history text to select important dates in the history of your civilization. You may also consult online sources.

A Game: What Date Is That? *(Cont'd.)*

Special Considerations

- Be sure the dates you select are accurate.
- Refer to Data Sheet 15.2 to learn how to write numbers using ancient number systems.
- Carefully recheck your numbers when written in an ancient number system. Because the dates on your number line will be converted to the Hindu-Arabic system, use B.C. if necessary. (Of course, before the birth of Christ, there was no "B.C." designation for dates.)
- Before drawing your time line, decide when it will begin and when it will end. Divide the time line into equal units, and place dates as close as you can to where they should appear on the time line.
- Use black felt-tipped or heavy black ballpoint pens to draw your time line. The black will reproduce clearly on a photocopy.
- When you draw your time line, do not label the dates with their events. Be sure to write down the answers on a separate sheet of paper.

To Be Submitted

1. Your time line with dates written in the ancient number system
2. A key with the dates in your number system and a list of events that they refer to

Notes

_____

_____

_____

_____

_____

_____

_____

_____

_____

Ancient Number Systems

We use the Hindu-Arabic number system today. It was developed in the lands of the Middle East and India before 800 A.D. and eventually spread throughout the Western world during the 1100s. As you will see when you examine the ancient number systems below, the Hindu-Arabic system was a major improvement.

Egyptians (about 3000 B.C.)

The ancient Egyptians wrote numbers the same way they wrote words: tediously, by carving or scratching pictures in stone or wood.

| = 1 (picture of a finger)

∩ = 10 (picture of a heel bone)

𝕁 = 100 (picture of a coiled rope)

𐦥 = 1,000 (picture of a lotus flower)

ſ = 10,000 (picture of a bent finger)

𓆓 = 100,000 (picture of a tadpole)

𓀠 = 1 million (picture of an astonished man)

To write their numbers, the Egyptians repeated the symbols, with the largest number on the left and other numbers in descending order.

The date 1290 B.C. (keeping in mind that the Egyptians would not have actually used the designation B.C.) was the year Ramses II became pharaoh of Egypt. He ruled for sixty-seven years during Egypt's Golden Age. In the ancient Egyptian number system, the date 1290 B.C. would be written in this way:

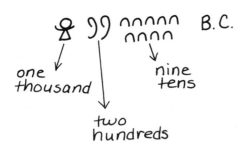

Ancient Number Systems *(Cont'd.)*

Greeks (about 100 B.C.)

The ancient Greeks were the first to use alphabetic symbols to stand for numbers. They were able to express large numbers by using twenty-seven symbols. They used their alphabet, which had twenty-four letters and borrowed the other three letters from other alphabets.

α = 1	⌐ = 7	μ = 40	ρ = 100	ψ = 700	͵ϩ = 4000
β = 2	η = 8	ν = 50	σ = 200	ω = 800	͵ε = 5000
γ = 3	θ = 9	ξ = 60	τ = 300	π = 900	͵ϛ = 6000
ϩ = 4	ι = 10	ο = 70	υ = 400	ϙ = 1000	͵⌐ = 7000
ε = 5	κ = 20	π = 80	φ = 500	͵β = 2000	͵η = 8000
ϛ = 6	λ = 30	ϙ = 90	χ = 600	͵γ = 3000	͵θ = 9000

The Greeks wrote numbers by placing the largest symbol on the left and the corresponding symbols for the other numbers to the right, in descending order.

The year 429 B.C. saw the birth of Plato. The ancient Greeks would have written the date as:

$$\underset{400}{\upsilon} \quad \underset{20}{\kappa} \quad \underset{9}{\theta} \quad B.C.$$

Romans (about 100 A.D.)

The ancient Romans could express all numbers from 1 to 1 million with a minimum of seven symbols:

I = 1	C = 100
V = 5	D = 500
X = 10	M = 1,000
L = 50	

They built numbers by putting symbols together in the following way:

- A smaller number to the right of a larger one means to add.
- A smaller number to the left of a larger number means to subtract.
- A vinculum, written as a — , placed over a letter means to multiply that value by 1,000.

It was on March 15, 44 B.C. (known as the Ides of March), that Julius Caesar was assassinated. The Romans would have written the year in this way:

$$\begin{array}{cc} X \; L & I \; V \quad \text{B.C.} \\ 50-10 & 5-1 \\ \text{or} & \text{or} \\ 40 & 4 \end{array}$$

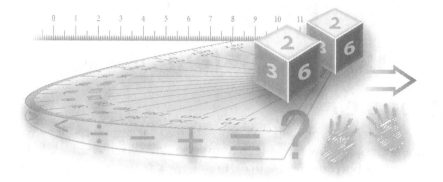

An interview to math's past

One way to view mathematics is to see it as a tool, a means of expressing quantities and values. As such a tool, math is reflected in a society's economy, customs, and routines. For example, in the 1930s, the cost for spending an afternoon in a movie theater was about a dime (often to see a double feature). Today the cost is much, much more and counting. The increase in ticket prices is an indication of change in our society. Change is frequently shown through a comparison of numbers. In this project, students will view the past through numbers, gaining an appreciation of how our society has changed over the years.

Goal

Working alone, students will interview a senior citizen, preferably a relative. They will ask this person what life was like when he or she was young, and particularly what role mathematics played in his or her life. After completing the interview, students will present their findings to the class in an oral report. *Suggested time:* Two to three class periods over a two-week period.

Math Skills to Highlight

1. Recognizing the changing role of mathematics as a society develops through time

2. Using technology in problem solving

Special Materials/Equipment

Pads or a clipboard for note taking; note cards for organizing oral presentations. *Optional:* A tape recorder for taping the interview; computer with Internet access for research.

Development

Students are used to learning about the past through books. Seldom do they learn about it through oral history, that is, speaking with someone who has lived it.

- Begin by explaining that students are to interview a senior citizen about what life was like, especially in terms of mathematics, when he or she was young. Students should try to interview a grandfather, grandmother, great-uncle, or great-aunt. A great-grandparent would be even better. If older relatives are unavailable, suggest that students interview older friends of the family. Discourage them from interviewing people they do not know well.

- Distribute copies of Student Guide 16.1. Review the information with your students, and make certain that they understand what they are to do. In particular, note the sample questions, which will help keep the interview focused on math.

- Distribute copies of Data Sheet 16.2, "Tips for Conducting Effective Interviews." Remind students that since they are looking for mathematical facts of the past, they must ask questions that will elicit responses about math. Suggest that students consult print and online sources to familiarize themselves with the time period their interview will cover.

- Mention that during the interview, students should look for ways that the use of math has changed. For example, clerks at cash registers today seldom figure out how much change to give to a customer after a purchase. The register computes the change for them. More examples: credit cards, debit cards, and ATM cards have become major means of transactions for many people. Still others do their banking using the computer. People shop with their computers as well, purchasing just about anything they wish from online stores.

- After completing their interviews, students should organize their notes to make an oral presentation of their findings to the class. They should use note cards for this.
- You may want to limit the length of time for the presentations. About three minutes is reasonable. Less makes some students hard-pressed to finish sharing their information; more adds up to a significant amount of class time.

Wrap-Up

Students present their findings orally to the class.

Extension

Working in groups, students brainstorm how the use of mathematics might change in the future. A spokesperson for each group shares the group's conclusions with the class.

STUDENT GUIDE 16.1

An Interview to Math's Past

Situation/Problem

To find out the role mathematics played in the past, you are to interview a senior citizen. After your interview, you will present your findings to the class in a short oral report.

Possible Strategies

1. Decide who you will interview. A grandfather, grandmother, great-uncle, or great-aunt will have spent his or her youth at a time when life was quite different from today. If you are unable to interview a relative, consider interviewing an older close friend of the family. Do not interview anyone without your parent or guardian's permission.

2. Determine in advance the time period your interview will focus on. This will help you ask questions that will provide you with information about the role math played in society at that time. Consult print and online sources to learn about the time period.

Special Considerations

- Since the focus of your interview will be on mathematics, your questions should zero in on math. Following are some areas you might like to explore with your interviewee:

  What was the pay per hour (or the salary) of your first job? What was your job, and what were your responsibilities?

  What was the price of a gallon of gasoline?

  What was the cost of a new car?

  What was the price of a new house? What was the cost of renting an apartment?

  What were the prices of clothing, shoes, and food?

  What did people do for entertainment when you were young? How much did this cost? What was considered a fun activity?

An Interview to Math's Past *(Cont'd.)*

What kind of math did you learn in school? How much math homework did you receive? What was your math class like? How did you figure out answers to long or difficult problems?

Did you ever need to use math on a job? If yes, in what way?

In what ways have you seen mathematics change during your life?

What surprises you about math's role in society now?

- Add your own questions to those listed above.
- Pay close attention to major events that might have influenced the life of your interviewee. The Great Depression and World War II, for example, affected countless people. If they did affect your interviewee, find out how.
- Refer to Data Sheet 16.2 for advice on how to conduct an effective interview.
- In organizing your notes, compare the math facts of the past to the uses of math today. Look for similarities and differences. Note any major changes.
- In preparing your oral presentation, use note cards to help you remember important facts. Be sure to number your cards.
- If your teacher puts a time limit on your presentation, make sure that your talk fits into the allotted time. Rehearse your report; time yourself if necessary.

To Be Submitted

Your note cards.

Rough Notes

_____

_____

_____

_____

_____

_____

_____

Tips for Conducting Effective Interviews

You can learn much valuable information about a topic during an interview. Following are some suggestions that will help you to get the most out of any interview you conduct.

- Think about who is best able to give you the information you are looking for.

- Learn as much about the subject in advance as you can. This will help you to formulate questions that will provide you with useful information.

- Think of questions before the interview, and write them down. Use questions that require explanations. Avoid questions that can be answered with a simple yes or no.

- Be ready to ask a follow-up question to clarify or expand an answer you receive.

- If you are not sure about an answer, ask for an explanation.

- Consider recording your interview on a tape recorder. Keep in mind that some people do not like to be recorded. Ask first. If they object, do not use a recorder. If they do not mind, be sure you have a clean tape and fresh batteries. (Take along a pen and pad just in case.)

- If you are taking notes with a pen and pad, do not try to write down everything your interviewee says. Focus on the main points. Use your own personal shorthand of abbreviations. For example, *John* is J, *math* is m, *and* is +. Inventing your own codes will help you to take notes quickly.

- For important facts or statements, try to write down the person's exact words. You can quote the person after the interview. Whenever writing a person's exact words, be sure to use quotation marks.

- After you have asked your questions, end the interview by thanking the interviewee for his or her time. Do not keep the interview running after it is done. Writing a thank-you note a few days later is a nice gesture.

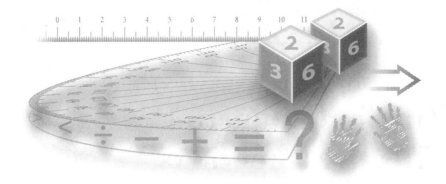

Rating consumer products

We live in a society where we have a choice of purchasing countless consumer products. Just consider how many types of sneakers are for sale or how many brands of potato chips we can choose from. To help us select the best product from among many, we can turn to articles and books that rate various products. Most rating systems rely on statistics, which are built on mathematics.

Goal

Working in pairs, students will select products to research. They will evaluate and compare the features and prices of each product and try to determine which is best for their needs. They will display their results using charts, tables, or illustrations and share their conclusions orally with the class. *Suggested time:* Two to three periods over two weeks.

Math Skills to Highlight

1. Gathering and analyzing data
2. Comparing prices and features of products
3. Making decisions based on collected data
4. Creating charts to illustrate conclusions
5. Using technology in problem solving

Special Materials/Equipment

White poster paper; rulers; markers; felt-tipped pens for making charts. Students will need to have access to the products they select to research; however, in most cases, they will either already have access to them or the products will be inexpensive and easy to obtain. Copies of articles or samples of magazines such as *Consumer Reports,* in which products are rated, will give students an idea of how products may be compared. *Optional:* Computers for consulting online sources for research; an overhead projector, PowerPoint, or interactive whiteboard for presentations.

Development

Ask your students what rationale they use to buy products. For example, why did they buy the sneakers they are wearing? Most likely, they tried on various styles, were impressed with the sneaker through advertising, or heard about it from a friend. They might have seen somebody wearing the sneaker and decided they liked it. Explain that people usually have reasons for buying one product instead of another.

- Begin the project by explaining that students will work with a partner. They will select a type of product (see Data Sheet 17.2, "Tips for Comparing Some Everyday Products," for some examples), and rate samples of the product produced by different companies. Ultimately, they are to determine which brand is best.

- This project will likely require meeting outside class. Since students who are friends usually find it easier to work together at home, try to allow friends to be partners.

- Distribute copies of Student Guide 17.1, and review the information with your students. Emphasize that they should compare at least three brands of the same product, although encourage them to compare as many brands as they can.

- Distribute copies of Data Sheet 17.2, and go over the material closely with your class. Mention that the products listed on this sheet are just some examples of the types of products they might decide to compare. There are many others. You might wish to brainstorm examples of more potential products with your class; list them on the board or an overhead projector. This will give students more choices. Direct students to inexpensive products and products to which they have easy access.

- Explain that since the features of different products vary, students should select features or categories for comparison based on their specific product. Video games would be compared differently from brands of pretzels, for example.

- Suggest that students may compare products by using them, tasting them (in the case of food), reading their specifications or ingredients, and asking others about them.

- After students have finished their research, they should write down their findings and reasons. They will present their conclusions to the class.

Wrap-Up

Students present their findings orally to the class using charts, tables, or illustrations to support findings. Consider having students use an overhead projector, PowerPoint, or interactive whiteboard in their presentations.

Extension

Suggest that students select a product they like, and encourage them to consult *Consumer Reports* or other publications to find out if the brand they use is the one recommended in terms of price and quality. Frequently students are surprised that the brand they use is overpriced or of lesser quality than competing brands according to their rating guide.

STUDENT GUIDE 17.1

Rating Consumer Products

Situation/Problem

You and your partner are to select a product, and compare at least three brands of it. You are to compare its features and price, and try to determine which is the best choice for your hard-earned dollars. After you have reached your conclusions, design charts, tables, or illustrations to support your findings, and present your data to the class in an oral report.

Possible Strategies

1. Consult examples of articles that compare or rate products. The magazine *Consumer Reports* is a good source. Copies may be in your school or public library. You may also consult online sources.

2. Brainstorm with your partner which types of products you might like to research. Consult Data Sheet 17.2 for some ideas.

3. Divide the tasks for the project. Although you will work together on much of this, you may find it helpful if one of you interviews people for research, while the other compares product specifications. If your partner is more artistic than you, he or she may design the chart and you handle the oral presentation.

Rating Consumer Products *(Cont'd.)*

Special Considerations

- After selecting your product, decide which features or qualities you will compare. Write these categories on a sheet of paper, then compare the products.

- Consider developing a rating scale. You might use something like 1—Superior, 2—Fair, 3—poor. Compare each brand according to your scale.

- Along with their physical features, products may also be compared according to their specifications or ingredients.

- You may find it helpful to ask friends and acquaintances about specific products. To ensure accurate results, always ask the same questions in the same manner. Write down the questions you intend to ask ahead of time. Focus your questions on features or qualities.

- After obtaining your data, analyze the information, and make decisions comparing qualities to price.

- Create charts, tables, or illustrations to highlight your results. Sketch rough copies of these first. This enables you to revise before starting the final copy. Arrange the design so that it provides the data clearly. List your products by brand name and your categories for comparison. Consider designing charts and tables on a computer.

- Before presenting your findings to the class, write notes so that you do not forget to mention any important information. Rehearse your presentation.

To Be Submitted

1. Research notes
2. Chart, tables, or illustrations

Name _____

Tips for Comparing Some Everyday Products

Popular products are compared and rated regularly in magazine articles and books. Many educated consumers rely on such articles to help them buy the products that fit their needs and budgets.

Some Ways Products Are Compared

Special features	Ease of use
Simple directions	Taste
Texture	Ingredients
Price	Comfort
Power	Durability
Resale value	Attractiveness

Some Products to Consider Comparing

Video games	Video game systems
Potato chips	Sodas
Pizzas	Fast-food restaurants
Ice cream	Internet service providers
Pretzels	Sneakers
Clothes	Computers and/or printers
Cell phones	Digital cameras
CDs	DVDs

Examples of Products and Features or Qualities to Compare

Video games: action, color, graphics, realism, sound, excitement, ease of use, easy-to-read manual, price

Pretzels: type, taste, saltiness, crispness, fat content, natural versus artificial flavoring, ingredients, cholesterol, price

Sneakers: purpose, comfort, durability, style, color, price, nonmarking soles, technology

Jeans: comfort, style, price, texture

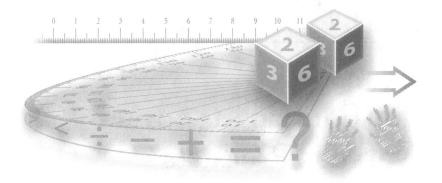

Lands of ethnic origin: A statistical potpourri

If your class is like most others, you probably have students who come from various backgrounds. Exploring the countries of students' ethnic origins can be an interesting activity. Asking students to investigate countries of origin from a mathematical perspective can impress them with the value of math, as well as help them to learn about the countries from where their families came.

Goal

Working individually, students will research a country of their ethnic origin and use mathematics to describe it. On completion of their research, students will create a mathematical fact sheet, a "statistical potpourri," highlighting information about their country. *Suggested time:* Two class periods over two weeks.

Math Skills to Highlight

1. Researching statistical information about selected countries
2. Analyzing and organizing data
3. Using mathematics to communicate ideas
4. Using technology in problem solving

Special Materials/Equipment

Reference books, including almanacs and atlases to research countries; white paper for the fact sheet; colored pencils; rulers; crayons; markers; felt-tipped pens; correction fluid. *Optional:* Computers and printers to design the fact sheets; Internet access for research.

Development

Discuss with your students that the United States is a country of immigrants. All of our families have roots elsewhere. Even the ancestors of Native Americans came to North America across a land bridge between Asia and Alaska during the Ice Ages.

- Begin this project by explaining that students will work alone and research a country of their ethnic origins. If students have several backgrounds in their family, they should pick one or, if they are ambitious, two or more. If a student is unsure of his background, give him the option of simply selecting a country that interests him.

- Note that students are to base their research on math in order to create a fact sheet of statistics.

- Distribute copies of Student Guide 18.1 and review the information with your class. Be certain that they understand what they are to do.

- Hand out copies of Data Sheet 18.2, "Tips for Creating a Country's Fact Profile Sheet." Review the material with your students, and point out that the categories listed are some types of facts they should look for during research. It is likely that some students will be able to find more, and some students will have trouble finding information about all the categories on this list. Obviously the extent of the information available depends on the country.

- To help students with their research, you may wish to take the class to the school library for one period. Inform the librarian in advance that you will need books about various countries. It would help to poll your students and write down the countries they will be researching. Particularly note any obscure ones, which will help your librarian provide the necessary materials. Encourage students to consult online sources as well. Note that students should continue their research outside class.

- After completing their research, students are to create a statistical fact sheet about their country.

- Remind them to arrange their data in a clear and logical manner. Suggest that they do a rough draft first. They should include their sources on the bottom of their fact sheet.

Wrap-Up

Display the students' fact sheets.

Extension

Suggest that students research and compare the statistical data they found on their country with the state they live in or one of the large states of the United States. California, Texas, New York, and Michigan are good examples. These states will be larger and have bigger populations and greater economies than many countries around the world. Discuss this with your class.

162

STUDENT GUIDE 18.1

Lands of Ethnic Origins: A Statistical Potpourri

Situation/Problem

You are to research the country of your ethnic origin from a mathematical perspective. When you are done with your research, you are to create a fact sheet highlighting the information you found.

Possible Strategies

1. If you are from several ethnic backgrounds, select the country you would most like to learn about.

2. Consult books and online sources that contain information about your country, and focus on information that is expressed in numbers. See Data Sheet 18.2 for examples.

Special Considerations

- Take accurate notes. List your sources on your notes. This is helpful should you need to recheck some facts. It is also helpful because you must include your sources on the fact sheet.

- After completing your research, make a rough copy of your fact sheet. Create a design that is easy to read and attractive. Consider how you would like to organize your information. You might organize your facts according to major categories, alphabetically, or numerically in order of size.

- Consider designing your fact sheet on a computer and printing the sheet. If you design the fact sheet by hand, use rulers to draw straight lines on the sheet.

- Use markers or felt-tipped pens for final copies. Correction fluid is useful if you make a mistake.

- You might want to include a drawing of your country's flag or a map of your country.

- Be sure to include your sources of information at the bottom of your fact sheet.

To Be Submitted

1. Notes
2. Fact sheet

163

Tips for Creating a Country's Fact Profile Sheet

The following categories are often expressed in mathematics. Note that these are just a sample. There are many others.

General Facts

- Population
- Number of states or provinces
- Number of major cities and their populations
- Largest and smallest cities
- Percentage of people living in cities
- Percentage of people living in specific regions
- Number of major newspapers
- Number of major religions
    - Percentage of people belonging to each religion
- Education
    - Number of elementary schools
    - Number of secondary schools
    - Number of universities
    - Number of school-age children
    - Literacy rate
- Currency (U.S. equivalent)
- Size of armed forces
    - Army, navy, air force
- Number of languages
    - Percentage of people speaking each
- Government
    - Number of major bodies
    - Number of members
- Political parties
    - Percentage of population belonging to each

Tips for Creating a Country's
Fact Profile Sheet *(Cont'd.)*

Geography

- Latitude and longitude
- Time zone
- Area
- General elevation
  - Highest mountain
  - Lowest place
- Longest river
- Largest lake
- Percentage of forested land
- Land use facts

Climate

- Number of climate zones
- Average annual and monthly temperature
- Average annual and monthly precipitation
- Average annual snowfall
- Average summer temperature
- Average winter temperature
- Noteworthy weather facts

Economy

- Gross national product
- Per capita income
- Employment rate
- Percentages of white-collar, manufacturing, and service occupations
- Average length of workweek
- Amount of national debt (or surplus)
- Imports
  - Total value and major types
- Exports
  - Total value and major types
- Percentage of land used for farming
- Energy
  - Types and amounts

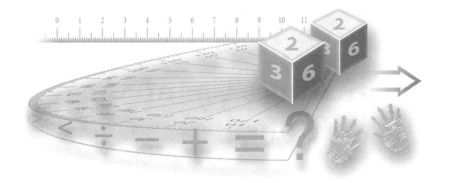

A land of immigrants

The United States is a land of immigrants. Even the ancestors of Native Americans crossed over the land bridge between Asia and Alaska during the Ice Ages. Unfortunately, although all of our ancestors have emigrated to America, newcomers have not always been welcomed by the general population. This is especially true today when many Americans feel that the number of immigrants permitted into the country should be reduced.

Goal

Working in groups of two or three, students will research immigration, identifying those decades in which the most immigrants came to the United States. They will also note from which countries the most immigrants came. Groups will then design graphs to illustrate their data. *Suggested time:* Three to four class periods.

Math Skills to Highlight

1. Researching mathematical data
2. Organizing data
3. Using percentages
4. Designing graphs to illustrate data
5. Using technology in problem solving

Special Materials/Equipment

Reference books on immigration; poster paper; felt-tipped pens; markers; rulers. *Optional:* Computers for designing graphs; printers; Internet access for research.

Development

This project is ideal for incorporating technology into your math curriculum. Students can consult online sources for research and create graphs of their findings with computers. This project also offers an excellent opportunity for collaboration with your students' social studies teacher, who may assume responsibility for research and discussion while you handle the analysis of information and graphing.

- Start the project by explaining to students that they will work in groups of two or three. They will research the history of immigration to the United States, identifying the years in which immigration was the greatest, and noting from which countries the immigrants came.

- Discuss the concept of the melting pot and that the United States is a land of immigrants.

- Explain the controversy over immigration today. Many Americans feel that immigration should be reduced. These people believe that immigrants take jobs from Americans and put a strain on social services. Although many facts about immigration are in dispute, that does not ease the arguments.

- Hand out copies of Student Guide 19.1, and review the information with your students. Especially emphasize the following:

    Groups will need to do several graphs for this project. Some will focus on the number of people who came to the United States during specific time periods, and others will show the countries from which most immigrants came.

Suggest that students design graphs according to decades. Trying to do each year individually will make constructing the graphs too tedious.

- Distribute copies of Data Sheet 19.2, "Researching Immigration," which will help students with their research, analysis, and design of graphs.

- To facilitate research, schedule a period for your class in the library. Ask the librarian in advance to reserve books that contain information on immigration. Suggest that students also consult online sources in their research.

- If you have access to computers, encourage students to enter their data on spreadsheets and from there create graphs to illustrate their information. Students may also create their graphs using poster paper and markers.

Wrap-Up

Display the graphs of your students.

Extension

Encourage students to examine the issue of immigration today. Suggest that they investigate the number of legal versus illegal aliens in the United States. What are the numbers? What are the implications, problems, and concerns? Discuss their findings with the class.

STUDENT GUIDE 19.1

A Land of Immigrants

Situation/Problem

Working with a partner or group, you are to research the history of immigration to the United States. Find the decades in which immigration was the greatest and also identify from which countries the most immigrants came. You are then to create graphs to illustrate your findings.

Possible Strategies

1. Divide the task of research. For example, in a group of three, one of you might investigate the years 1789 to 1869, another the years 1870 to 1939, and another from 1940 to the present.
2. Discuss which type of graphs will best present the information you find. It is likely that you will need to create several graphs.

Special Considerations

- Take accurate notes, and maintain a list of your sources. This is important if you need to go back and check information. Consult both print and online sources.
- Do not try to record immigration for every year since colonial days. Rather, focus on decades from 1789 (the year of the first meeting of the U.S. Congress) to the present.
- Use Data Sheet 19.2, "Researching Immigration," to help you with your research, organize your information, and select your graphs.
- Use computers to create spreadsheets and generate graphs.
- If you use markers and pens for creating your graphs, use pencils and rulers to make light lines on your paper so that your graphs and letters will be straight. Use colors that stand out, and highlight your information. Be sure to label your information.

To Be Submitted

Your graphs.

Name _____

Researching Immigration

To help you with your research about immigration, use the following guidelines. Use separate sheets of paper for your notes:

- Find the total number of immigrants for each decade. Start with 1789–1799, then do 1800–1809, 1810–1819, 1820–1829, and so on, to the present.

- List the seven countries from which the greatest number of immigrants came. Remember that people have come from all around the world to start new lives in the United States. Compare the numbers of immigrants from these countries to the overall number of immigrants. Express these numbers as percentages.

- After gathering your data, determine how you will illustrate your information with graphs. Here are some suggestions:

    Bar graphs are useful for comparing years or decades.

    Line graphs are a good choice to show how immigration increased or decreased over a specific period.

    Circle graphs (pie graphs) are useful for showing percentages of immigrants and the countries from which they came.

    Pictographs use symbols to represent data, and offer a good visual perspective for comparison. Values are approximations.

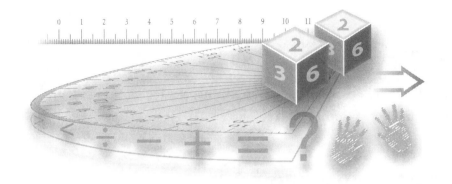

The games people play

Mathematics is relevant to people everywhere for work and play. Too often students think of mathematics in association with work. They seldom pause to consider how math is found in virtually every game we play, from the most basic application (keeping score), to strategies based on mathematical probabilities in games such as chess, poker, and dice. Math is a vital element of games that have originated around the world. In this project, students will learn about games that involve math and have originated in various countries or regions. This activity could be an excellent addition to a multicultural event in your school or may simply be an enjoyable activity of its own.

Goal

Working as partners, students will select a game that originated in a country other than the United States, research its development, and write a short historical report about it. They should emphasize how mathematics is necessary for the game. They will then teach and play the game with at least two other students. *Suggested time:* Two to three class periods.

Math Skills to Highlight

1. Math skills will vary according to the game, strategies, and scoring.
2. Using technology in problem solving.

Special Materials/Equipment

Since students will either obtain or make the games on their own, materials will vary and should be supplied by students. *Optional:* Computers with Internet access for research.

Development

Ask students what they do for recreation. You will get a lot of different answers: watch TV, play or watch sports, play video games or board games, solve puzzles, go to the movies, dance, participate in crafts or art, or simply hang out. Emphasize that there are plenty of recreational activities from which to choose. Prior to the advance of technology, however, there were far fewer choices. In the past, people often turned to board and card games to relax and pass time. Games were played almost everywhere, and the most popular were carried across borders and spread around the world.

- Begin this project by explaining that for this project, students, working with a partner, will select a game that originated in another country. They will either obtain the game or make a sample copy of it, research its history, and share their findings with the class. They will also have the opportunity to teach and play the game with other students.
- Distribute copies of Student Guide 20.1, and go over the information with your students. Emphasize that students should not go out and buy an expensive version of the game they select. (If it is inexpensive, they may wish to purchase it, however. Some games, for example, need nothing more than a deck of cards.) If the game is expensive or they cannot find a version of it, they should make their own, using poster paper, cardboard, and markers.
- Hand out copies of Data Sheet 20.2, "Games from Around the World," which provides a list of possible games students might select. Note that this is only a partial list; encourage students to conduct research to identify other games they might wish to use for this project.
- Discourage students from choosing games that require physical activity. These will be difficult to manage in the classroom.
- Once students have decided on a game, they should research its history using both print and online sources. Many games were invented because

of unusual circumstances. Solitaire, for instance, was created in the eighteenth century by a Frenchman who had been imprisoned and sentenced to solitary confinement. He had plenty of time in which to think of things to do. (Today there are several versions of solitaire, some with different names.)

- Caution students that various sources may report different rules or a different history for the same game. For some games, it might be hard to pinpoint a single country of origin. Instruct students to pick one version of the game and record their source.

- When they complete their research, partners are to write a brief historical background of their game and provide a sample of the game for the class. They will explain the history of the game, its rules, and its mathematics components, and they will play a sample round with other students. Set aside a period to play games.

- *A note on evaluation:* Rather than collecting all the games, you might instead circulate around the classroom as students are playing the games. If you need to grade the games, doing it now eliminates the need for collecting them and handing them back. (This is certainly a consideration for games that have a number of small pieces.)

Wrap-Up

Students play their games in class.

Extension

Students may wish to pursue other contributions to mathematics from other countries, such as origami, tangrams, or tessellations.

STUDENT GUIDE 20.1

The Games People Play

Situation/Problem

You and a partner are to select a game that involves math and originated in another country. You are to write a historical background of the game, emphasizing how mathematics is important to the game, obtain or make the game, and teach others how to play it.

Possible Strategies

1. Select a game that you have played that you know was invented in another country.

2. Choose a country or a region of the world in which you are interested, and research a game that originated there.

3. Choose a game from Data Sheet 20.2. Or research and choose another game.

Special Considerations

• Look at Data Sheet 20.2, and carefully go over the list of possible games. Choosing one with which you are familiar will make your research easier.

• Choose board games, card games, or pen-and-pencil games. Avoid games that require physical action.

• Conduct research, using both print and online sources.

• Be aware that many games have several variations. Choose one variation, and record your source.

• During your research, you may find that the country of origin for your game is in dispute. Try to find at least two sources that agree, and use the information they provide. If all of your sources provide different information, choose one. Be sure to record the source.

174

The Games People Play *(Cont'd.)*

- As you research, look for the following information:

   History of the game

   Object of the game

   Number of players

   Types of playing pieces

   Rules

   How math is involved in the game

   Any special evidence of the culture from which the game evolved

- After you have finished your research, write a short historical report about your game. Be sure to include the importance of math.

- Obtain or make a sample of your game. If you make it, create a board (if necessary), playing pieces, spinners, dice, and other parts. Include the object of the game and a copy of written rules.

- To avoid losing pieces, store game pieces in a shoe box, a plastic bag, or similar container.

- Become the expert on your game. With your partner, explain it to some of your classmates and show them how to play.

To Be Submitted

Historical report.

Notes

_____

_____

_____

_____

_____

_____

_____

DATA SHEET 20.2

Games from Around the World

Following are some popular games and the country or region of their origin:

Achi—Ghana
Alquerque—Arabia
Backgammon—Mesopotamia
Checkers—Europe
Derrah—Nigeria
Dominoes—Egypt
Draughts—France
Dreidel—Germany (although the game has been a part of Jewish culture)
Go—Central Asia and Japan
Goose Game—France
Hala tafl (Fox Game)—Scandinavia
Konane—Hawaii
Ko-no—China and Korea
Lu lu—Polynesia
Magic Squares—China, Egypt, India, Western Europe
Mu Torere—New Zealand
Nim—China
Nine Men's Morris (also known as Mill, Morelles, and Muhle)—Europe
Ninuki-Renju—Japan
Nyout—Korea

Paiute Walnut Shell Game—Paiute Indians, Nevada
Parcheesi—India
Patolli—Aztecs (Mexico)
Pipopipette (also known as Boxes)—France
Pong Hau K'i—China
Seega—Egypt
Senet—Egypt
Sey—West Africa
Shut the Box—France
Sir Tommy—England
Snakes and Ladders—India
Solitaire—France
Tac Tix—Denmark
Tapatan—Philippines
Totolospi—Hopi Indians, southwestern United States
Wari—West Africa
Who Crosses the River First?—China
Yote—West Africa

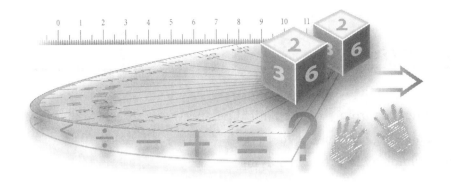

Back to the past

Mathematics has been a part of virtually every civilization throughout history. From the most primitive times when people tallied numbers by making slashes on cave walls, to the vast empires of the ancient Romans and Chinese, and now to our modern world linked by computers, people have needed numbers for counting and calculating. Although the need for math has not changed, its methods certainly have. People have always adapted mathematics to fit their needs.

Goal

Working in groups of three or four, students will select a country or region and research how mathematics was used in the past. They will present their information to the class in oral reports. *Suggested time:* Two to three class periods over two weeks.

Math Skills to Highlight

1. Researching mathematical data
2. Organizing data from a mathematical perspective
3. Communicating ideas about math
4. Using technology in problem solving

Special Materials/Equipment

History and other reference books; note cards. *Optional:* Computers with Internet access for research; overhead projector and transparencies, PowerPoint, and interactive whiteboard for presentations.

Development

Consider working with your students' social studies or history teacher for this project. While you may permit students to research countries or regions anywhere in the world, most students will find it more relevant to research places they have studied in their social studies or history classes.

- Begin this project by explaining that students will work in groups of three or four. They are to select a country or region, and research its use of mathematics in the past. They may choose to research a specific period, for example, the Middle Ages, or they may decide to trace the use of math from ancient times to the present.

- Hand out copies of Student Guide 21.1, and review the information with your students.

- Distribute copies of Data Sheet, 21.2, "An Outline to Organization," and go over the material with your students. Especially point out the areas on which they might concentrate their research, but note that these are merely suggestions. Groups may find many other ways math was used in the places they are researching. This data sheet also offers a plan to organize the information they find.

- Emphasize the importance of dividing the task of research for this project. Some topics may require quite a bit of research. If group members consult different sources, they will be able to collect information more efficiently than if they consult the same sources.

- To help students with their research, reserve a period in the library. Confer with the librarian in advance, and ask him to set aside books that will help students with their research efforts. Encourage students to conduct research on their own time and suggest that they also check online sources for information.

178

Project 21

- Note that research should focus on how mathematics was used in the past.
- Remind students to take accurate notes and record their sources.
- After students have completed their research, provide some class time for the groups to organize their information. Suggest that they design support materials such as charts, tables, and illustrations to enhance their presentations. Recommend that they use file cards to write down notes they will use in their presentation.
- For the oral presentations, we suggest you limit each group to about three minutes.

Wrap-Up

Students will make oral presentations. Consider having students use an overhead projector, PowerPoint, or an interactive whiteboard in their presentations.

Extension

Students may write a report on their topic and compile their reports in a math history anthology. Some students may wish to make a table of contents and index. This would be a great exercise in organization and cross-referencing.

Name _____

Back to the Past

Situation/Problem

Your group is to select a country or region of the world, and research how mathematics was used there in the past. You may pick a specific time period, or you may investigate the use of math from ancient times to the present. After you have finished your research, you will present your findings orally to the class.

Possible Strategies

1. Brainstorm with your group to determine which country or region you would like to research. Selecting one that you have already studied (or are studying) in social studies or history class will make your research easier, because you will be somewhat familiar with your topic.

2. After deciding on a country or region, select a time period. Again, it might be best to choose one you are studying or have already studied.

3. Divide the task of research. Some countries may require more research than others. By having each member of your group consult different reference books, your group will be more likely to uncover information. Consult online sources as well.

Special Considerations

- Focus your research efforts on the use of mathematics in the time period you have chosen.

- Accurately record your sources.

- After you have completed your research, work together to organize your information. See Data Sheet 21.2 for some suggestions.

- Create charts, tables, or illustrations to support your information.

- Select a spokesperson or persons to present your oral report. Use file cards to write notes.

- Rehearse your presentation so that you may share your findings clearly and smoothly.

To Be Submitted

1. Notes for your presentation
2. A list of sources of your information

DATA SHEET 21.2

An Outline to Organization

The development of mathematics has progressed through the centuries. In researching the use of math in a specific country or region in the past, consider the following topics:

Math in Trade and Commerce

Math and Farming

Math and Money

Math in Education

Math in Science

Math in Weights and Measures

Math and the Number System

Math in Time

Math and Calendars

Math in Exploration

Math and Inventions

Math in Architecture

Math in Recreation and Sports

Computation Machines or Tools

Math in Art

Math in Music

While there are many ways to organize reports, two of the most effective are the *chronological pattern* and the *logical pattern:*

Chronological pattern: Ideas are arranged in the order in which they occurred. What happens first appears first in the report, what happens second appears second, and so on. This organizational method works best when ideas follow in sequence.

Logical pattern: Ideas are grouped according to a plan. For example, related ideas may be grouped together. Ideas may be arranged from least important to most important or most important to least important.

Section Three

Math and Language

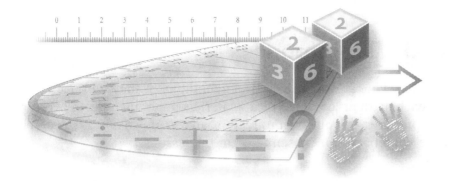

Becoming the experts

Few students appreciate the role and responsibilities of their teachers, yet many would welcome the opportunity to teach their classmates. This project offers them the chance by allowing students to assume the role of the teacher and teach a math lesson to the rest of the class. Placed in this role, students are required to develop, organize, and articulate their thoughts on a topic in math, thus becoming experts on it.

Goal

Students will work in groups of four or five and present a math lesson to the class. *Suggested time:* Three to four class periods.

Math Skills to Highlight

1. Using math as a means to communicate ideas. Specific math skills will vary depending on the topic and content of the lesson that individual groups select.

2. Using technology in problem solving.

Special Materials/Equipment

Students should provide a list of materials and equipment they will need a few days prior to teaching their lesson. Typical materials might include an overhead projector, manipulatives, calculators, protractors, and so on. *Optional:* PowerPoint and interactive whiteboard for presentation of the lessons.

Development

Before introducing this project, decide if you want the students to focus on particular topics or subjects. For example, if your current unit of study is geometry, you might provide a list of topics or let students develop their own topics but only in geometry. Of course, you can leave the scope of this project open and permit students to select a subject or topic of their own choosing.

- Begin this project by explaining to your students that they will become experts and teach the class a math lesson. Tell them which topics they may cover or if they have the freedom to select a topic of their choice.
- Emphasize the importance of researching, developing, organizing, and planning their topic.
- Distribute copies of Student Guide 22.1, and review the information with your students. Suggest that they divide tasks equally among group members.
- While you should provide a class period for students to work in groups and prepare their lessons, encourage students to meet and work outside class if necessary. Depending on the lessons your groups present, it is likely that you will need at least two class periods for the groups to teach their lessons.
- Hand out copies of Data Sheet 22.2, "A Student Sample Lesson Plan Guide," and discuss the material with your students. Explain that this is a guide similar to ones that teachers use to develop and present lessons. Following it will help students develop an effective lesson. We recommend that groups try to limit their lessons to about ten minutes.
- Hand out copies of Worksheet 22.3, "A Sample Lesson Plan," and suggest that students use the sample as they plan their lesson. Review the worksheets before students present their lessons to the class. Offer suggestions for improvement, but allow students to have the final decisions for their plans.
- Make sure that you obtain the materials students will need to present their lessons. Securing materials in advance reduces the chances for oversights.
- Encourage students to practice their lesson ahead of time.
- Schedule the lessons, and assign each group a date that they are to present their lesson.

Wrap-Up

Students present their lessons to the class.

Extension

Invite students to present their lessons to students in other classes.

STUDENT GUIDE 22.1

Becoming the Experts

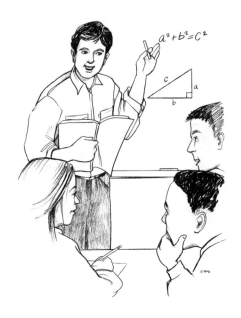

Situation/Problem

Your group will plan and teach a math lesson to your class.

Possible Strategies

1. Brainstorm with your group to decide on a lesson that you would like to teach.
2. Divide tasks among group members to make the project easier to manage.

Special Considerations

- Because there are various tasks to researching and developing a lesson, divide them. For example, individual group members may be designated to handle some of the following:

  Research

  Making visual aids (models, charts, posters, transparencies, graphs)

  Writing the plan

  Obtaining and using equipment necessary to your lesson

  Teaching the actual lesson

  Preparing activities for students (sample problems, worksheets, quizzes)

Becoming the Experts *(Cont'd.)*

- Review Data Sheet 22.2, and use the information to guide you in developing your material.

- Complete Worksheet 22.3, which will help you to produce a good lesson. Submit the worksheet to your teacher for approval.

- Select who will present the lesson to the class. You may decide on one group member or have two, three, or everyone take a part.

- Practice and coordinate the presentation of your lesson. Make certain that you have any special materials or equipment you need ahead of time. Make sure you know how to work the equipment and that the equipment works.

- When presenting the lesson, speak clearly, and be willing to answer questions.

- If you distribute any handouts to your classmates, make sure that they understand what they are for and what they are to do with them. All directions should be clear.

To Be Submitted

A copy of your group's lesson plan.

Notes

_____

_____

_____

_____

_____

_____

_____

_____

_____

_____

_____

DATA SHEET 22.2

A Student Sample Lesson Plan Guide

There are many kinds of lesson plan formats that teachers may use. Following is a common one that you can use in planning your math lesson:

Objective: What do you want the students to know after you have taught the lesson? What will they be able to do?

Method: How are you going to teach the lesson? Some suggestions include:

Pose a problem.

Demonstrate using a model.

Show examples on the board or an overhead projector.

Illustrate the concept on a poster.

Make a table, chart, or graph.

Use calculators or computers.

Provide practice problems.

Procedure: What are you going to do first? Second? Third?

Materials: What do you need to assemble or prepare in advance? What supplies or equipment will you need?

Evaluation: How will you know if you accomplished what you set out to do? In other words, how can you show that students learned what you taught?

A Sample Lesson Plan

Topic: _____

Objective:

Method:

Procedure:

Materials:

Evaluation:

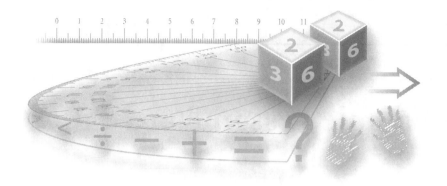

Great debates

A debate in math class? Absolutely. Debates, which are usually associated with English and social studies classes, are a unique and enormously beneficial activity for math students. To prepare for a debate, students must research their topics, clarify their thinking, and formulate arguments. They must understand topics thoroughly and be in command of the facts. Such things fit in perfectly with a mathematician's skills. Moreover, because there is an element of competition in debates, they appeal to students, most of whom find debates enjoyable.

Goal

Working in groups of four or five, students will prepare and engage in debates about issues, problems, and topics in math. *Suggested time:* Four to five class periods spread over two to three weeks.

Math Skills to Highlight

1. Skills will vary according to the material of each debate. Accuracy of facts and arguments should be emphasized.

2. Using technology in problem solving.

192

Special Materials/Equipment

Reference books on a variety of issues and topics in math; two long rectangular tables students may use during the actual debates (however, you can push desks together). *Optional:* Computers with Internet access for research.

Development

There are many formats for debates, ranging between the formal and informal. For this project, we suggest you use an informal format (see Data Sheet 23.2, "An Informal Debate Structure"), which will be easier to manage.

- Begin this project by explaining to your students that they will take part in debates on problems or issues in math.
- There are many possible topics for math debates. Here are some:

  Should calculators be used in the primary grades?

  Are standardized test scores a good indication of student achievement?

  Should students help their teachers decide what to study in math?

  Should homework be counted in the evaluation of students?

  Should all students have the opportunity to take honors or accelerated math courses?

  Is learning mathematics necessary for all students?

  Should the history of mathematics be studied in the typical math class?

  Is it still necessary to learn multiplication tables?

  Should parents help their children with math homework?

- If you wish, you may generate debate topics with your class. Conduct a brainstorming session, and ask students to suggest topics. As they do, write them on the board or an overhead projector. Do not take the time to expand or focus them now; simply generate as many possible topics as you can. Afterward, go back and eliminate those that will not lend themselves to a debate, and focus on those that will.
- Although you can have all groups debate the same topic, it is usually better to give each set of debate teams its own topics. This allows more issues to be addressed and adds variety to the arguments. If a group feels strongly about a topic, let them debate it. Few things add more to a debate than passion.
- When you organize your students into teams, be sure to mix abilities. For example, if your school has a debate team and some members of the class are members, divide them among your groups. Allowing seasoned debaters to be on the same team will stack the odds. If a team has one fewer member than its opponent, a member of that team may speak twice during the debate.

Great debates

- Hand out copies of Student Guide 23.1. Review the information with your students, and emphasize the suggestions for conducting effective debates.

- If you decide to use the suggested debate format, distribute copies of Data Sheet 23.2. Although there are many debate formats, this one works well with most classes. You may adapt it to your class. Note that the times provided are flexible and may be changed to fit your needs.

- If possible, reserve library time to get students started with research. Depending on the topics, you will probably need at least one period for this. Instruct students to continue research on their own, consulting both print and online sources.

- After students have completed their research, give them a class period to work with their teams and organize their material. Remind them to refer to their student guides. Be sure that students understand the debate format and their roles.

- Long tables provide an excellent atmosphere for a debate. If you cannot obtain tables, have students push their desks together at the front of the room.

- You should serve as the moderator and timekeeper. It is probable that you will have to remind some students of the debate process.

- Although scoring is an important part of formal debates, we recommend avoiding it. Many students will feel undue pressure if their arguments are being evaluated. Concentrate the activity on the debate of topics related to math.

- Requiring students to hand in their notes and sources is a good way to monitor that each group member contributed to the project.

Wrap-Up

Stage the debates. Depending on how long the debates are, this may take three or four class periods. You may schedule the debates over time so that they do not disrupt your normal schedule.

Extension

Videotape the debates, and make the tape available to others. (Be sure to check the policies of your school before videotaping any student. The policies of some schools prohibit photographing or taping students; others require release forms signed by both students and parents or guardians.) Students will enjoy watching themselves in the debate process.

194

STUDENT GUIDE 23.1

Great Debates

Situation/Problem

Your group will participate in a debate about a problem or issue in math.

Possible Strategies

1. Once you receive your topic and position, analyze it carefully so that group members understand it.
2. Divide the tasks of research equally among group members.
3. Formulate ideas and arguments to support your position.

Great Debates *(Cont'd.)*

Special Considerations

- Be accurate in your note taking. Focus your research on evidence that will support your position. Such evidence often comes from:

    Authorities and experts

    Studies

    Government or business statistics or research

    Documented facts

    Personal experience

- Consult both print and online sources.

- Record all your sources in case you need to recheck some information. Be sure your facts are accurate.

- As you research and find support for your position, be alert to facts that may support your opposing team's position. Noting them will help you to respond to their arguments.

- After gathering your information, analyze it and identify the strongest points that support your position. Build your arguments around these points.

- Decide which members of your group will present which points. The first speaker, who will present your opening statement, should be one of your team's better speakers.

- Be familiar with the debate format you will follow. Each member should know when and on what he or she will speak. (If you are not sure of the debate format, check with your teacher.) In case a team member is absent on the day of the debate, have another member ready to take his or her place.

- Before the debate, practice your presentation. Other members of the team should feel free to offer advice. Remember that comments should be positive and helpful.

- During the debate, take notes of your opponent's arguments and facts. Be ready to rebut (or argue against) them as the debate continues.

- When debating, speak clearly, and use facts. Support your statements with evidence. Keep track of the time, and do not get sidetracked on minor issues.

- In preparing to rebut your opponent's arguments, look for mistakes in facts, opinions not based on evidence, weakness in examples, or weakness in logic.

To Be Submitted

Copies of your notes and sources.

An Informal Debate Structure

A debate involves taking sides on an issue and then presenting your arguments for or against that issue. A statement of that issue is called a *proposition.* The people who support the proposition are called the *affirmative team;* they agree with the statement. Those who disagree with it are called the *negative team.* An *opening statement* introduces a team's position and offers important evidence. A *rebuttal* is a team's response to its opponent's arguments. A *second statement* is a team's chance to expand on their ideas and evidence.

While formal debates follow a strict format, informal debates have various structures. Following is an example of a format for an informal debate. Each statement is usually given by a different team member:

Opening statement by affirmative team—3 minutes
Opening statement by negative team—3 minutes

2 minutes allotted for teams to prepare rebuttals

Rebuttal by affirmative team—2 minutes
Rebuttal by negative team—2 minutes

Second affirmative statement—3 minutes
Second negative statement—3 minutes

2 minutes allotted for teams to prepare rebuttals and closing statements

Rebuttal and closing by affirmative team—2 minutes
Rebuttal and closing by negative team—2 minutes

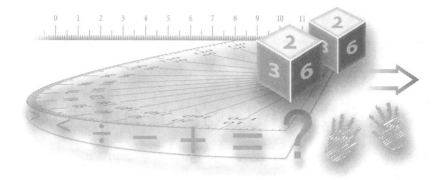

The mathematics publishing company

Creating, designing, and publishing a mathematics magazine can be an excellent project for a math class. Class magazines provide students with an opportunity to work together on various topics in math in ways that move beyond the typical classroom setting, especially in the integration of mathematics and writing. Although producing a math magazine may sound like a lot of work, it is not as much as you may think, and the rewards are significant. High-quality magazines can be produced using computers, printers, and photocopiers.

Goal

Working in groups of four to six, students will create and produce a mathematics magazine. *Suggested time:* Four to five class periods, although this will vary based on the length of the magazine. Work time on the magazine may be divided over several partial class periods.

Math Skills to Highlight

1. Using writing as a method to communicate ideas about math
2. Using technology in problem solving

Special Materials/Equipment

Tape; scissors; glue; standard printing and copying paper; clip art; rulers; computers; printers; photocopier. *Optional:* Internet access for research and posting the magazine online.

Development

This is a fine interdisciplinary project, and you may wish to work with your students' English teacher, who can help students with the actual writing of the magazine, while you handle the math and production. If this is your first experience in doing a class magazine, we suggest that you limit its size. Four pages (two sides back to back) will be long enough for several articles and short enough to manage easily.

- Before beginning the project, establish guidelines regarding the content of the magazine. You might make such decisions yourself or discuss them with your students. All of the material should have a math slant. You might choose to do a general magazine or concentrate on a particular topic such as geometry, fractions, or integers.

- Start the project by explaining to your students what they will be doing. If you have copies of magazines that other classes produced, distribute them so that students can see what a math magazine can be. It is likely that many of your students have never been involved in the production of a class magazine. If you wish, you might brainstorm possible names for the magazine.

- Distribute copies of Student Guide 24.1, and discuss the information with your class, highlighting what you believe are its most important points.

- Assigning different responsibilities to the groups divides the work and makes it easier for you to coordinate and oversee students' efforts. Group 1, for example, might focus on writing articles, Group 2 might be responsible for news items, while Group 3 handles design, layout, and artwork. Group 4 may be responsible for puzzles, and Group 5 might create tricky math problems. Consider arranging the groups according to student interests. Artistic students will probably prefer to work on design and illustrations, and those who like to write will most likely want to handle the articles. To ensure that groups work efficiently, assign a leader for each group whose responsibility is to keep the group on task.

- To support your students in their writing, hand out copies of Data Sheet 24.2, "Possible Ideas for a Math Magazine," which offers suggestions for possible articles. Emphasize that these are just some ideas and that students might come up with many more.

- You may wish to distribute and discuss copies of "The Writing Process and Math" in Chapter Two of Part One. The information on the handout may be helpful to students as they develop, research, and write their material.

- Encourage your students to write articles on computers. Saving articles electronically facilitates revision.

- Artwork and graphics may be added to your magazine through computer software, clip art, or line drawings. Most word processing programs include clip art. Art can also be downloaded from other electronic sources and imported to your file. *Caution:* Use only clip art that is copyright free. If students use hand-drawn illustrations, they should be kept simple and dark for photocopying. Complex drawings with heavy shading do not reproduce well on copiers. Avoid having students draw on a finished page. Instead, illustrations should be drawn on a separate page, cut out, and pasted or taped onto a finished page. This reduces the chances of ruining the page during illustrating. (The art teacher can be a good source of materials and advice.)

- With many activities going on simultaneously, it will be necessary to set deadlines to ensure that work gets done on time. Set reasonable deadlines, and stick to them.

- When all the work is finished, have students proofread the material once more to catch any final errors or oversights. (You may find it necessary to proofread the material as well.) The magazine can then be photocopied. Photocopying on both sides of a page not only results in a more attractive magazine but also saves paper.

Wrap-Up

Publish enough copies of the magazine so that each student gets one and there are enough left over for distribution to other classes, administrators, and display in the school library. You might also publish your magazine on a school or class Web site.

Extensions

Make this an ongoing project. Produce additional magazines throughout the year. If you teach several math classes, you might have each class produce a magazine at a different time. Consider having students submit material to a schoolwide math magazine.

STUDENT GUIDE 24.1

The Mathematics Publishing Company

Situation/Problem

This class will be divided into groups, and work together to create and produce a mathematics magazine. Each group will be responsible for a different part of the magazine. Upon completion, copies of the magazine will be printed and distributed.

Possible Strategies

1. Break your group's tasks down into small parts, and divide the parts among group members.
2. Communication is important among the groups to avoid duplication of effort. Groups should keep each other informed of their progress.

Special Considerations

- Generate ideas for articles by brainstorming with your group.
- Some topics may require research. Any research should be thorough.
- You will need to establish deadlines for material to be completed. Managing time is important.
- Try to finish articles well in advance of deadlines. This will give you time for editing and revision. Expect things to go wrong.
- Use computers to write and print your material. Desktop publishing software can produce outstanding magazines.

The Mathematics Publishing Company *(Cont'd.)*

- If your magazine will include artwork, you will need to decide on what types. Possibilities include clip art or line illustrations drawn by students. Many word processing programs contain clip art, which can be easily placed in articles. Clip art can also be obtained from online sources or from art and drawing programs. Be sure any art you use is copyright free.

- If you are writing math problems, puzzles, or games, be sure to include an answer key.

- Design your magazine. Create a "dummy" layout where you place copies of the articles on a white background sheet that serves as the magazine page. Arrange articles and artwork attractively. Use white tape or paste to attach pieces. Many software programs can do this for you.

- Before printing, proofread the magazine a final time. Special care should be given to any math problems that appear in the magazine. Accuracy is important.

To Be Submitted

Your group's contribution to the magazine in its final form.

Notes

_____

_____

_____

_____

_____

_____

_____

_____

Name _____

DATA SHEET 24.2

Possible Ideas for a Math Magazine

Following are some suggestions for articles that might appear in a mathematics magazine:

General articles about math

Historical pieces

Biographical sketches of famous mathematicians

Women in mathematics

Information about math contests

Updates on the school math team

Pieces about student math projects

Self-help articles

Features about class mathematicians

Math shortcuts

Articles on measurement

Articles on estimating

Articles about money

Articles on math and science

Short articles about the math club

Games

Puzzles

Math Problem of the Month

Tricky math problems

Articles on careers in math

Reviews of math software

Guides for purchasing home computers

Math trivia

Math cartoons

Examples of Possible Titles

Informational article—"Math in Our Lives"

Self-help feature—"How to Study for Math Tests"

Biographical sketch—"Karl Friedrich Gauss: Boy Genius"

Article about money—"Budget Tips for Teens"

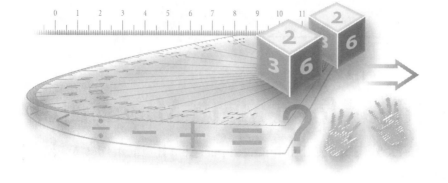

Rating math Web sites

The Internet offers an enormous amount of information for math students and teachers. Some Web sites provide excellent material, some provide good material, and some provide material that might be unclear, or, worse, inaccurate.

Goal

Working in pairs or groups of three, students are to review at least three Web sites devoted to mathematics and rank them according to specific criteria. They are to write a report in which they describe the sites and then compare and rank them. They will also share their findings and conclusions with the class in an oral presentation. *Suggested time:* Two to three class periods.

Math Skills to Highlight

1. Gathering and organizing data
2. Comparing and evaluating mathematical information
3. Using technology in problem solving

Special Materials/Equipment

Computers with Internet access. *Optional:* Overhead projectors and transparencies, PowerPoint, and interactive whiteboards for presentations.

Development

Ask your students to think about Web sites they have visited. Ask them what distinguishes a good site from a mediocre one. Explain that for this project, they will visit and rank math Web sites.

- Begin by explaining that students will work in pairs or groups of three. Each group will visit and evaluate at least three Web sites devoted to math and rank them according to five criteria: design, ease of navigation, interesting content, accuracy, and degree of interaction (if any). (*Interaction* refers to a Web site's functionality that enables a visitor to play a game, find the answers to questions, or interact with the Web site in some other manner.)
- Distribute copies of Student Guide 25.1, and review the information with your students. Emphasize that they are free to choose the Web sites they evaluate and suggest they visit several before making their decision.
- Hand out copies of Data Sheet 25.2, "Tips for Rating Math Web Sites," and discuss the information with your students. The sheet offers useful tips for finding and evaluating Web sites.
- If your school has a computer room with Internet access, arrange to take your class there so that students will be able to work together under your supervision. Depending on your students, this might require one or two periods, especially if you want them to complete their research in school. Otherwise, instruct students to work outside class to complete the project.
- Remind students that they are to write a report briefly describing the individual Web sites they selected, comparing them, and ranking them.
- Suggest that they create a chart to show their rankings.
- Remind them that they must present an oral presentation on their conclusions. Consider having students use overhead projectors, PowerPoint, and interactive whiteboards in their presentations. An excellent method would be to project Web sites on the whiteboard during the presentations.

Wrap-Up

Display the reports and rankings.

Extension

Have groups visit the Web sites that other groups rated. Then determine the class's choices of the top three math Web sites.

STUDENT GUIDE 25.1

Rating Math Web Sites

Situation/Problem

You and your partner or group is to select three Web sites devoted to math, evaluate them, and rank them according to these criteria: design, ease of navigation, interesting content, accuracy, and degree of interaction. (*Interaction* refers to whether a Web site allows a visitor to the site to play a game, ask questions, search for information, and so on.) You will summarize your evaluation in a report that includes the rankings in the form of a chart and share your conclusions with the class in an oral presentation.

Possible Strategies

1. Visit and evaluate several Web sites. Compare and contrast the features of three of the sites.

2. Record or mark the URLs (addresses of the Web sites) so that you may return to them easily.

3. Discuss with your partner or group the features of the various sites.

Rating Math Web Sites *(Cont'd.)*

Special Considerations

- Keep accurate notes as you evaluate each Web site.
- Determine a ranking system. You may simply rank sites from 1 to 3 on each criterion, with 1 being the best, or you may designate rankings with symbols, for example, a star. Whatever system you decide on, it should be clear and consistent.
- Refer to Data Sheet 25.2, "Tips for Rating Math Web Sites," for help in finding and evaluating sites.
- If necessary, set up a time when you and your partner or group can work together in researching Web sites. Viewing sites individually may make it difficult to discuss and rank them.
- When writing your report, use an opening, body, and conclusion. Be sure to include reasons for your rankings.
- Display your rankings of the Web sites as a chart. Consider designing your chart on a computer.
- For your oral presentation, include examples of features of the Web sites that support your conclusions.

To Be Submitted

1. Your report with rankings
2. Any charts and supporting materials

Notes

_____

_____

_____

_____

_____

_____

_____

DATA SHEET 25.2

Tips for Rating Math Web Sites

The following information can help you find and evaluate Web sites that offer information about math:

Finding Sites

- Go to Yahooligans at www.yahooligans.com. Under School Bell, click on Math, and math Web sites of interest to students will be displayed.

- Use other search engines and the term "math for students" to find additional Web sites about math.

- Pay close attention to links, which can lead to still more sites.

Rating Sites

- Select the criteria you will use to evaluate the Web sites.

- Visit several sites before choosing those you will evaluate.

- Evaluate each site according to your criteria. Compare and contrast the features of each.

- Be sure you can support your conclusions. If you and your partner or group disagree about features of a particular site, seek a compromise.

- Rank the sites in a consistent manner. Use numbers or symbols for ranking.

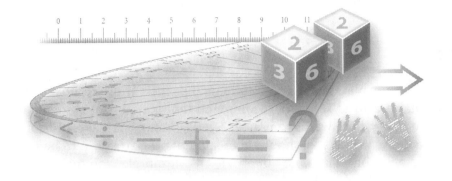

Fictional numbers: Writing a story

Students have been writing stories in their English classes since the time they could hold a pencil. But how many of your students have ever had the chance to write a story in math? This project offers them the opportunity, along with allowing their creativity to fire up.

Goal

Each student will write a story in which mathematics plays a major role. *Suggested time:* Two class periods.

Math Skills to Highlight

1. Writing as a means to express ideas about math
2. Using technology in problem solving

Special Materials/Equipment

A few dictionaries and thesauruses. *Optional:* Computers and printers; Internet access for research (if necessary).

Development

Begin this project by explaining to your students that they are to write a story in which math has a central role. We recommend that you permit students to write any type of story they wish: romance, science fiction, fantasy, mystery, historical, or comedy. While you should give students the freedom to be creative, you should also consider some boundaries. For example, you may prefer that they do not write stories of gory horror, or graphic or suggestive sex, or that have foul language. Here are some other suggestions.

- Two class periods should be enough for this project. A Monday and Thursday or Tuesday and Friday will give students a chance to work on their stories at home as well as in school.
- Emphasize that mathematics should be an important part of the stories.
- If you have some students who are reluctant writers, consider allowing them to work with a partner and be coauthors. This will reduce anxiety.
- Hand out copies of Student Guide 26.1, and review the information with your students. In particular, emphasize the elements of a story's plot as noted on the student guide. Although some of your students are probably familiar with a basic plot structure from their English classes, others will need to be reminded. Understanding the components of a plot will help students in creating their stories. Also, be sure to mention the sample story ideas. They can give students a boost in starting.
- Distribute copies of Data Sheet 26.2, "A Revision Checklist for Stories." Discuss the material with your students, pointing out that the sheet provides valuable suggestions for revising stories.
- Encourage students to use sound writing skills; they should edit and revise their work. You may wish to hand out copies of "The Writing Process and Math" in Chapter Two of Part One.
- Encourage students to use computers in the writing of their stories. This will make the work of revision easier.

Wrap-Up

Display the finished stories on a bulletin board. You might also set aside some time for class readings in which students swap and read each other's stories.

Extensions

Compile stories in a class anthology. Brainstorm a title with your students, photocopy and bind the stories, and distribute them around the school. If you find a story that is suitable for reading to younger children, arrange an opportunity for the author to visit students in that class and read his or her story to them.

STUDENT GUIDE 26.1

Fictional Numbers: Writing a Story

Situation/Problem

You are to write a story in which mathematics has a major part in the plot.

Possible Strategies

1. List as many potential ideas for stories as you can.
2. Select the story idea you like best and expand it, adding characters, action, and details.

Fictional Numbers: Writing a Story *(Cont'd.)*

Special Considerations

- When creating your story, keep in mind these important parts of a plot:

    A lead character has a problem that he or she must solve.

    Conflict, in the form of hostility, anger, or resentment, is associated with the problem. In trying to solve the problem, the character may come into conflict with other characters, nature, or himself.

    The lead character makes plans to solve the problem, but each plan fails and the problem grows worse. These setbacks are called complications.

    The lead character keeps running into complications, and the problem keeps getting worse until the story reaches a climax. At this point the character either solves the problem and triumphs, or he fails to solve it.

    Most stories have either a happy or sad ending, depending on whether the character has solved the problem. Some stories conclude with unresolved endings in which the character neither succeeds nor fails.

- Remember to make math a major part of the story. Be creative. Here are some examples of story ideas:

    Inspector Integer solves his biggest case in which a computer genius electronically empties people's bank accounts.

    A future team of space explorers is mapping the distance to a nearby star where alien life is thought to exist. Their goal: first contact.

    Your lead character needs to come up with cash fast. The story shows why and how.

    A thief in the school breaks into lockers. Your character discovers a pattern to the thefts: the thief chooses lockers according to a special sequence of numbers.

    Your character, who is a numbers whiz, thinks that winning the math contest will be easy. But solving the puzzle for the grand prize proves to be harder than she thinks.

- Use dialogue and action in your story.
- Describe scenes with vivid details.
- Write your story on a computer; this will make revision easier.
- Edit and revise your story so that your finished copy is an example of your best writing.

To Be Submitted

A copy of your finished story.

DATA SHEET 26.2

A Revision Checklist for Stories

The following list can be helpful when you are revising your story. Apply the questions below to your writing.

1. Does my story make sense? Is it realistic and believable?

2. Do my characters act like real people? Do they speak like real people? Do they dress like individuals, according to their natures? Have I described them so that my readers can picture them clearly?

3. Do my characters behave according to their natures? Are their actions logical?

4. Does my story have a problem that the characters must solve? Is the problem big enough to hold the interest of readers?

5. Does my story have conflict?

6. Have I described my scenes with color, sounds, and other details? Do my scenes paint pictures in the minds of my readers?

7. Does every scene in my story build to the climax? Is my climax exciting?

8. Have I used periods, commas, and other punctuation correctly? Have I used quotation marks to show dialogue?

9. Have I used correct spelling? Have I used words in their proper contexts?

10. Am I satisfied that this story is the best I can make it?

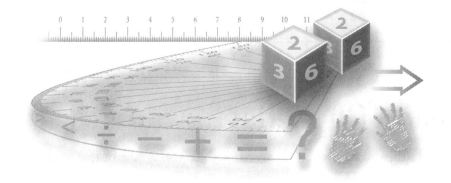

A mathematical autobiography

Our lives are interwoven with numbers. We use numbers to mark significant events (birthdays, for example), as a means of identification (social security numbers), and as a way of ranking and comparison (most students prefer a test average of 98 in math over an average of 68). Numbers help us to track the condition of the economy and tally the score in sports. They tell us our height, weight, and waist size. Most students accept numbers as a part of life without realizing their true significance. This project may help them to understand just how much they rely on numbers.

Goal

Working individually, students will write a mathematical autobiography. *Suggested time:* Two class periods.

Math Skills to Highlight

1. Writing as a means by which to examine and express ideas about math

2. Using technology in problem solving

Special Materials/Equipment

A few dictionaries and thesauruses. *Optional:* Computers and printers for writing and revising.

Development

Begin this project by explaining that an autobiography is an account written by an individual about his or her life. Most students should be familiar with examples of autobiographies, and you might ask them to volunteer titles of some they have read. An autobiography might run several hundred pages and cover everything from a person's birth to old age, or it may be only a few paragraphs or pages focusing on an important aspect of the author's life. In that case, it is usually referred to as an autobiographical sketch. For this project, most students will be writing sketches.

- Hand out copies of Student Guide 27.1, and review the information with your students. Answer questions they may have. It is important to reduce any anxiety they may feel about writing.

- Instruct students to concentrate on the ways numbers play a part in their lives. If a student has experienced a major event in which numbers were important, he or she may wish to focus the writing on that topic. (Most students will find it easier to write about a specific topic or event than writing generally about numbers.) For example, one ninth-grade boy wrote about his best Little League baseball season. He managed the highest batting average in the league, led the league in home runs, and had the most wins and strikeouts as a pitcher. He noted that it was the only season he played that well, and he will always remember "the numbers." Another good example is a piece about a health program in which a student may keep track of workout time, repetition of specific exercises, nutrition, and the caloric content of foods. All involve numbers.

- Distribute copies of Data Sheet 27.2, "Organizing Your Autobiography." While there are many ways students may organize their material, the plan suggested on this sheet is simple and should make the task of writing easier.

- Consider giving students a chance at the beginning of the project to discuss the project with a partner or small group. This will help generate ideas for writing.

- Allow enough time for generating ideas, organizing the material, writing, and revising. For most students, two class periods should be enough. If you decide to set aside two periods, we recommend that you reserve a Monday and Thursday or Tuesday and Friday. Spacing the days will give students time to work on the project at home.

- Encourage students to use sound writing skills; they should edit and revise their drafts to polish their work. You may wish to hand out copies of "The Writing Process and Math" in Chapter Two of Part One.
- Encourage students to use computers when writing their autobiographies. This will make the task of writing and revision easier.

Wrap-Up

Display the finished autobiographies in the classroom or on a hallway bulletin board.

Extensions

Compile the autobiographies in a class book. Photocopy them, and make them available to others. Display the collection of autobiographies in the school library.

STUDENT GUIDE 27.1

A Mathematical Autobiography

Situation/Problem

You are to write a mathematical autobiography—a personal account of the importance of numbers in your life.

Possible Strategies

1. List various ways numbers affect your life. Take a sheet of paper and write down every way numbers are important to you. Do not worry about expanding any of the ideas now; just write down as many as you can.

2. Identify the most important ways numbers are important to you. Rank them in importance from most to least significant.

3. Select the most important way numbers affect your life. Expand your ideas on this topic, and use them as the focus of your writing.

Special Considerations

- In listing the ways numbers are important to you, think about:

Special dates	Exercise
Sports	Nutrition
Grades	Recreation
Money	Vacations
Height, weight	Goals, ambitions

  These are only some categories to consider. There are likely to be many more ways numbers are important to you.

A Mathematical Autobiography *(Cont'd.)*

- Organize your information logically; then use good writing skills to express your ideas.
- Write your autobiography on a computer, which will make revision easier.
- Be sure to revise your writing before submitting a final copy.

To Be Submitted

A final copy of your mathematical autobiography.

Notes

_____

_____

_____

_____

_____

_____

_____

_____

_____

_____

_____

_____

Name _____

Organizing Your Autobiography

- Most nonfiction writing follows a three-part plan: opening, body, and conclusion. Using this plan will help you to organize the information you have gathered for your mathematical autobiography.

- Your opening should:

  Capture the interest of the reader by introducing your subject. Try using an interesting or surprising statistic, state a problem in which numbers played a role, offer a joke or anecdote, or exaggerate a common situation in which numbers are important.

  Lead smoothly into the body of your autobiography.

- Your body should relate ways that numbers are important to you. Use the five W's to include details about events:

  What happened?

  Who was involved?

  When did the event happen?

  Where did it happen?

  Why did it happen?

  Add a "How." How do numbers fit in with what you are writing about?

- Your conclusion should:

  Contain a final idea for the reader to consider.

  Briefly tie the ideas of the piece together.

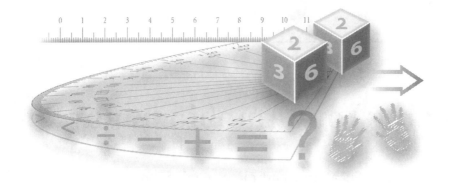

Puzzling

Puzzles can make any kind of learning fun. Most students enjoy solving math puzzles, particularly those created by their peers. Puzzles may serve as an introduction to new material, provide practice in reinforcing skills, and review concepts previously introduced. With a little guidance, students can design math puzzles of their own to share with their friends. They may even wish to publish their puzzles in math magazines or place them in their portfolios.

Goal

Students will work individually or in pairs to create math puzzles to share with the class. *Suggested time:* Two class periods.

Math Skills to Highlight

1. Specific skills will vary according to the types and content of puzzles. Care should be taken to ensure that puzzles and their answer keys are accurate.
2. Using technology in problem solving.

Special Materials/Equipment

Graph paper; rulers; correction fluid; non-repro blue pencils (a non-repro blue pencil does not show up on photocopied pages); black felt-tipped pens. *Optional:* Calculators; computers and printers that may be used in the design of puzzles; Internet access for research.

Development

When you introduce this project, many students may be at a loss as to how to create a math puzzle. A few days before starting the project, you might mention that the class will create mathematical puzzles and suggest that students consult math puzzle books, their texts, or online sources for ideas. Your school or local library will likely have many sources in the recreational math section. If you have examples of math puzzles created by students from other classes or examples from books, make them available. Data Sheet 28.2, "Puzzles, Puzzles, Puzzles," provides some examples, but there are many more. Also, prior to beginning the project, decide if you want students to concentrate on a particular unit of study or topic, or if they will be permitted to create puzzles on any topic they wish.

- Start this project by explaining what students will do. Tell them that puzzles should be designed with a particular purpose or objective. They may be used to introduce new material, review previously learned skills, or practice computation. Some ideas for the content of puzzles include:

  Learning definitions or properties

  Identifying figures

  Learning relationships between definitions and numbers

  Practicing computation or using calculators

  Solving proportions

  Applying order of operation rules

  Using percentages

  Finding perimeter, area, or volume

  Solving equations
- Distribute copies of Student Guide 28.1, and review the information with your students. Note that graph paper can be used to draw boxes for puzzles.
- Inform students if you want them to concentrate on specific topics.
- Distribute copies of Data Sheet 28.2, "Puzzles, Puzzles, Puzzles," which contains examples of puzzles students might do. Go over the sheet with your students, but emphasize that these are only some of the many types of puzzles they may create.

- Mention that students can add twists to common puzzles. A good example is a simple cross-number puzzle. Although the puzzle may focus on the basic operations of addition, subtraction, multiplication, and division, a new wrinkle would be to have the puzzle solvers use calculators and work against a time limit. Another idea would be for students to compete against each other, with a champion being determined by the fastest time to finding a solution. Such a puzzle would be fun and exciting, and give students practice in working with calculators.
- Suggest that students design their puzzles on computers. Various types of software may be used for drawing and writing. Many word processing programs come with easy-to-manage clip art, as well as possess drawing capabilities. Most include geometrical figures that can be easily manipulated on the screen.
- Suggest that students create a rough or dummy version of their puzzles before attempting to complete them.
- Emphasize that all math must be accurate and that each puzzle should have an answer key.

Wrap-Up

Make copies of the puzzles, and allow students to work on them. This may be done as a class activity on completion of the project.

Extension

Compile copies of the puzzles in a class book. Make the puzzle book available to other students.

STUDENT GUIDE 28.1

Puzzling

Situation/Problem

You will create a math puzzle that other members of your class will try to solve.

Possible Strategies

1. Study examples of math puzzles in magazines and textbooks, and on math Web sites. Pay close attention to see how they are created.

2. Determine your purpose or objective, and decide which type of puzzle you want to do.

Special Considerations

• Gather the math facts that you will use. Be sure all your information is accurate.

• Create a dummy version of your puzzle. Carefully sketch out on graph paper or a blank sheet how your puzzle will be set up.

• If you are using a blank sheet, use rulers to divide distances equally.

• If you need to draw boxes or spaces, use a fine-tipped black felt pen. This will reproduce well.

• Use a non-repro blue pencil to make marks that you do not want to be reproduced on a copier. This is useful for answer keys.

• Mistakes can be corrected using correction fluid.

Puzzling *(Cont'd.)*

- Consider creating your puzzle on a computer. Word finds, word scrambles, and tricky questions may easily be done on computers. Some software enables you to draw and arrange geometrical shapes.

- Double-check your work by exchanging your puzzle with a friend. He checks yours while you check his.

To Be Submitted

1. A final copy of your puzzle
2. An answer key

Rough Ideas

Name _____

Puzzles, Puzzles, Puzzles

Math puzzles come in countless forms. Here are a few examples.

Word Find

Pick mathematical terms, and arrange the words horizontally, vertically, backward, forward, and diagonally.

Find all the math terms.

E	R	A	U	Q	S
A	V	R	B	C	U
E	A	E	C	F	M
D	O	R	N	E	X

Solution

E	R	A	U	Q	S
A	V	R	B	C	U
E	A	E	C	F	M
D	O	R	N	E	X

Magic Square

The sums of the numbers in each column, row, and diagonal of the square are equal.

Find the missing numbers.

9	2	
4		8
5	10	

Solution

9	2	⑦
4	⑥	8
5	10	③

Cross-Number Puzzle

These puzzles are similar to crossword puzzles, except that digits are used.

ACROSS

1. The perimeter of a square whose side is 7
3. $3 \times 6 + 3$

DOWN

1. The area of a square whose side is 5
2. 9^2
3. The volume of a cube whose side is 3.

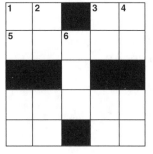

Solution

Puzzles, Puzzles, Puzzles *(Cont'd.)*

A Maze of Basic Facts and Computation

Solving problems helps you to find your way through the maze.

Use the solutions to the problems to guide you through the maze.

$5^2 - 1$

$2 \times 5 + 2^3$

$- 7 + 7$

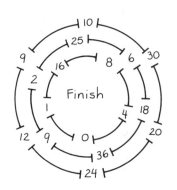

Solution

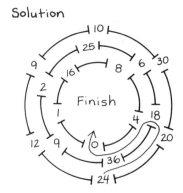

Mathogram

Letters are substituted for the digits of numbers in this puzzle. The object is to find the digit represented by the letter.

Find what digit each letter represents.

$$\begin{array}{r} TWO \\ + TWO \\ \hline FOUR \end{array}$$

Solution

$$\begin{array}{r} 928 \\ + 928 \\ \hline 1856 \end{array}$$

T = 9
W = 2
O = 8
F = 1
U = 5
R = 6

Word Scramble

The letters of math words are mixed up.

Unscramble each word to spell a math term.

enev

remip

Solution

even

prime

226

Puzzles, Puzzles, Puzzles *(Cont'd.)*

Rebus

A math term is written with words, numbers, and illustrations. The goal is to figure out what the term or sentence is.

A a ▬ = 2

Solution

A diameter equals two radii.

Secret Message

Answers to a list of math problems provide a secret message. Possible answers (in numbers) are paired with letters of the alphabet. By finding the correct mathematical answer, the correct letter is discovered. Once all the problems are solved correctly, the message is revealed.

Solve each problem. Match the answer with the letter of the alphabet to reveal a secret message.

C. $(6 + 3) \times 2 =$ _____
P. $(3 \times 5) - 1 =$ _____
A. $4 \times 3 =$ _____ $\times 4$
E. $200\% =$ _____

I. $4^2 =$ _____
L. $8 \div \frac{1}{4} =$ _____
R. $3.4 + 4.6 =$ _____
S. $5\frac{1}{4} - \frac{1}{4} =$ _____
Y. $3^3 =$ _____

$\overline{}$ $\overline{}$ $\overline{}$ $\overline{}$ $\overline{}$ $\overline{}$ $\overline{}$ $\overline{}$ $\overline{}$ $\overline{}$ $\overline{}$ $\overline{}$ $\overline{}$ $\overline{}$ $\overline{}$ $\overline{}$ $\overline{}$ $\overline{}$.
18 16 8 18 32 2 5   3 8 2   2 3 5 27   3 5   14 16

Solution

C. $(6 + 3) \times 2 = 18$
P. $(3 \times 5) - 1 = 14$
A. $4 \times 3 = 3 \times 4$
E. $200\% = 2$

I. $4^2 = 16$
L. $8 \div \frac{1}{4} = 32$
R. $3.4 + 4.6 = 8$
S. $5\frac{1}{4} - \frac{1}{4} = 5$
Y. $3^3 = 27$

Circles are easy as pi.

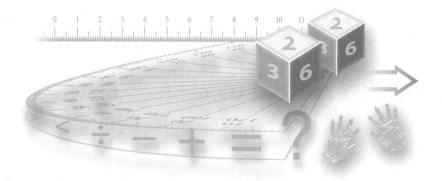

Lending a math hand

Students teaching students is a great learning activity with many benefits, especially when older students teach younger ones. Before they can present material to younger students, older students need to master their topics and articulate them in a manner that younger students will understand. Older students gain satisfaction (and often a boost in self-confidence) from helping younger children, while the younger ones learn math from a novel approach. Younger students are often more receptive to instruction from older students. Although there are many details to manage with this project, the overall benefits are well worth the effort required.

Goal

Working in groups of three or four, students will select a topic in math appropriate for younger students. They will write about it, creating a booklet that explains both concepts and applications, and then present their booklet to the younger students. *Suggested time:* Two to three class periods; additional time will be necessary to present the booklets to the students in other classes.

Math Skills to Highlight

1. Writing as a method of expressing ideas about math. Specific skills will vary, depending on the topic.

2. Using technology in problem solving.

Special Materials/Equipment

Black pens for line illustrations, examples, or practice worksheets; scissors and glue for cutting and pasting; clip art; staplers and staples for binding. *Optional:* Calculators; computers and printers for writing and printing their booklets.

Development

Before beginning this project, speak with teachers of students in lower grades in your district, and ask if they would like to have groups of your math students write and present booklets focused on specific topics to their classes. We suggest that you consider students in grades 1 through 5, for these are often the most receptive to visitors from other classes. Find out what types of material their teachers might want your students to cover. If possible, get a list of potential topics from each teacher. In the typical fifth grade, for example, basic operations are covered, along with fractions, decimals, and percentages. Once other teachers have agreed to have your students visit, set up tentative time frames. In planning visits, be sure to obtain any necessary permission from building principals, supervisors, or other administrators, as well as any necessary signed permission forms from parents or guardians.

- Introduce this project by explaining to your students that you have arranged the opportunity for groups to create math booklets that describe concepts and applications. They will write the booklets for students in lower grades (mention the grades). Once the booklets are finished, they will be photocopied and stapled. Your groups will then visit the classes they have written their booklets for and present the booklets to the younger students.

- If you have several grades and topics, you might permit your groups to choose which one they would like to do. If you have several teachers who have volunteered their classes, try to have a group write a booklet for each one. (If people volunteer and are left out, they may not volunteer next time.)

- To give your students an idea of what material for younger students is like, obtain some sample textbooks or workbooks from teachers of younger students and share them with your class.

- Distribute copies of Student Guide 29.1, and review the information with your students. Encourage students to refer to their guides as necessary throughout the project.

- Hand out copies of Data Sheet 29.2, "Creating a Math Booklet," and go over the material with your students. Note that they are to choose a specific topic and write about it, including concepts and applications. They may include practice sheets and examples. Point out that four sheets of paper will provide a booklet with eight pages (which includes the front and back covers). That should be enough for most groups.

- Suggest that students use computers for writing their booklets. Not only will revisions be easier, but many word processing programs come with clip art and features that can help in the design of booklets.

- Review the rough drafts of the booklets, and offer suggestions for improvement before the final copies are completed.

- Have enough copies of the booklets photocopied so that everyone in the classes that are to be visited will receive a copy.

- Encourage your groups to rehearse their presentation. The actual presentation may be little more than talking about the math topic, handing out the booklets, and going through them, or it may be an actual lesson that uses the booklet as a type of text. The final decision here should be based on the needs and abilities of the class your group is visiting, as well as the ability of your students, your schedule, and time constraints.

- Check with the teachers whose classes your students will miss and confirm that your students may miss class.

- Check your schedule, and make travel plans well in advance. If your students will need to travel to an elementary school across town, arrange for buses, costs, chaperones (if necessary), and permission slips with plenty of lead time.

- Check back with the teachers whose classes you will be visiting and confirm meeting times and topics.

- If possible, arrange for your groups to visit the same school (for different classes) on the same day. Having them make their presentations at about the same time reduces the logistics concerns. You can get one group started in one classroom and then head to the next.

Wrap-Up

Students present their math booklets to other classes.

Extension

Set up a tutoring program in which your students can help younger students with their math on a regular basis.

STUDENT GUIDE 29.1

Lending a Math Hand

Situation/Problem

Your group will select a math topic that is applicable to younger students. You will write a booklet that explains the topic; then you will present copies of your booklet to the students in their classroom.

Possible Strategies

1. Review with your group possible math topics you want to develop in a booklet.
2. Divide the tasks of the project to make the work easier to manage.

Special Considerations

- Select material that is appropriate for the age group you will be working with.
- Think about dividing tasks according to research, writing, layout, illustrations, and any practice worksheets that will be needed.

Lending a Math Hand *(Cont'd.)*

- Make sure you write clearly and explain your topic fully. Use language that your audience will understand. *Hint:* Check some books at the library that are written for the grade level of your audience. This will give you an idea of the vocabulary you should use. Material for second graders, for instance, is very different from material for sixth graders.

- Consider teaching a math concept in story form. Young children often learn easily through stories.

- Keep any illustrations and practice sheets simple. If possible, use clip art to highlight your work. If you use hand-drawn illustrations, be sure to use distinct lines. Avoid complex drawings, shading, or vivid colors, which may not reproduce well on copiers.

- Use computers to write your material. Because a booklet is set up differently from a page on a computer screen, you will either have to arrange your writing in columns (many word processing programs can do this for you), or write on only half the screen. If your software contains clip art, consider inserting art to illustrate your text. After printing your material, cut out the written text, and paste or insert it on the pages of your booklet.

- Refer to Data Sheet 29.2, "Creating a Math Booklet," for suggestions on how to produce your booklet.

- Practice how you will present your booklet to younger students. There are many possibilities. One group member may be selected as the lead presenter, while other members may act as assistants. If you divide the class, each group member may work with a group of younger students.

- When you travel to other classes, be sure you have all of your materials.

- When you present your booklets to other students, always be polite, considerate, and patient.

To Be Submitted

A final copy of your booklet.

DATA SHEET 29.2

Creating a Math Booklet

Following are suggestions for creating a math booklet that highlights math skills for younger students:

- Select a topic that is appropriate for the age and curriculum of the students who make up your target audience.

- Focus your topic. It is impractical to want to teach fifth graders fractions because the topic is far too broad. Try something more focused, like simplifying fractions or multiplying mixed numbers.

- Include concepts as well as steps to do the math in your booklet.

- Write steps to performing mathematical procedures in order.

- Provide examples for the steps.

- An easy way to make a booklet is to take two sheets of blank paper. Then follow these instructions:

  Put the sheets of paper together, one on top of the other.

  Staple along the edge of one of the longer sides (a staple on the top, the bottom, and in the middle).

  To help design the contents of your booklet, make a dummy layout. Think of this as practice. Experiment with where you might put writing, illustrations, and sample problems. You might need to try several possible layouts before you find the design you like the best. Make sure to number the pages correctly.

  If you want a longer booklet, add more sheets of paper.

- Leave ample space on each page. A page that is too crowded will not be attractive or easy to read.

- Add illustrations or clip art to your booklet.

- Design an attractive cover.

- Include practice worksheets or sample problems in your booklet.

- Proofread your work carefully, and make sure that all the examples of math are accurate.

- After your booklet is in its finished form but before you have stapled it, photocopy the separate pages. Remember to photocopy the pages of your booklet on the front and back. You will then need to staple the pages to produce the booklets. Be sure to make enough copies so that you, the teachers, and each person of your audience may have one.

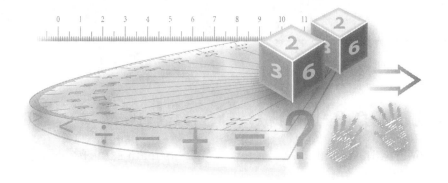

Sharing the math word

Having an article published can be a thoroughly satisfying experience for a student (or even an adult). Although math students and their teachers generally do not think much about publishing articles, there are many magazines and newsletters that are open to student writing on numerous topics, including math. That is the focus of this project: encouraging students to write articles about mathematics with the aim of publication.

Goal

Working individually (although you might allow students to work in pairs as coauthors), students will write articles on topics in math and submit them for publication. *Suggested time:* One class period to get the project started. Students will then work at their own pace, doing most of the work on articles at home.

Math Skills to Highlight

1. Writing as a means by which to express ideas about math. Specific skills will vary, depending on the topic of the article.

2. Using technology in problem solving.

Special Materials/Equipment

Computers and printers; photocopiers; regular business envelopes for letters; 9- by 12-inch mailing envelopes for student manuscripts; postage stamps. *Optional:* Internet access for research (if necessary) and checking Web sites for possible sources for publication.

Development

Since a major part of this project is writing, you may wish to work with your students' English teacher. He or she can work with the students on the writing while you focus your efforts on the mathematics. Due to the strong competition for getting an article published, you might want to make this an optional project only for students who indicate a strong desire to write and submit their work.

Many magazines and newsletters regularly publish student writing. Start looking for potential markets in your school. If your school board or PTA publishes a magazine or newsletter for parents or the community, see if the editors are interested in your students' submitting math articles. Perhaps they would be willing to run a regular column of student work.

Local newspapers and magazines are also potential markets. Contact the editors through a letter or a telephone call, and ask if they are interested in having your students submit material. Older students who are serious about writing should be referred to *Writer's Market* (Cincinnati, Ohio: F&W Publications, updated yearly). This source offers a listing of several thousand potential markets for writing. Your local library may have copies, or you may contact the publisher. You might also search the Internet for markets, using terms such as "publishing student writing."

- Begin this project by telling your students that although writing for publication is usually associated with English classes, potential markets are as open to well-written math articles as they are to articles written on other subjects.

- Note that submitting writing to magazines and newsletters is an excellent way to share ideas.

- Distribute copies of Student Guide 30.1, and review the information with your students. Note the importance of closely following the suggestions for submission.

- Recommend that students obtain sample copies of magazines they are interested in submitting articles to and also to search the Web sites of magazines for guidelines for the submission of articles.

- Distribute copies of Data Sheet 30.2, "A Sample Query Letter." Explain that a query is a letter an author sends to an editor of a magazine in which the author describes her idea for an article. In the letter, the author asks the editor if she would be interested in receiving a copy of the article. Queries save time and effort. If the author does not receive a positive response to a query letter, it saves her the effort of writing an article for which there is little interest.

- Discuss the query letter in detail. Explain that queries are written as business letters. Note the headings, greeting, and closing. Next, point out that query letters have distinct parts. Most begin with a statement of a problem or a direct question to the editor. In the sample, the statement is in the first paragraph. Some details are then offered (paragraph 2). After the details, the author asks the editor if she would be interested in the article and usually offers some detail about how the article would be structured (paragraph 3). Finally, the author offers his qualifications: why he is suited to write this article.

- Suggest that students use computers for writing their articles. Computers make writing and revision easier. Also, most editors frown on handwritten material.

- To help students generate ideas for articles, you might let them work in small groups for part of the period.

- If you assign this project to all of your students and wish to collect articles at a specific time so that you may review them, set a deadline.

Wrap-Up

Students submit their articles to magazines and newsletters.

Extension

Encourage students to write additional articles and submit them for publication. Make writing articles an ongoing activity in your class.

236

Sharing the Math Word

Situation/Problem

You are to write an article on a topic in math and submit it to a magazine for publication.

Possible Strategies

1. Identify magazines you would like to write for. Check their Web sites for their guidelines for writers.

2. Obtain copies of several magazines or newsletters that publish material written by students. Review the articles they publish, noting the types, lengths, and styles. Some magazines will send free sample copies if you request them.

3. Generate a list of potential topics, and narrow the list down to a topic you would like to write about. (Articles that are sent to magazines are called *manuscripts*.)

Sharing the Math Word *(Cont'd.)*

Special Considerations

- Try to select a topic that is fresh; in other words, it has not been done before. Or you might write about an old topic in a new way.

- Query magazines to see if they would be interested in your article idea. See Data Sheet 30.2, "A Sample Query Letter," for an example of a query letter.

- Always follow the magazine's guidelines. If the guidelines specify articles of 800 words, do not send an article with 1,200 words.

- When submitting a manuscript to a magazine, always include an SASE (self-addressed, stamped envelope). Magazines that reject your manuscript will not return your article without an SASE.

- Be sure the copy of your article that you send is clear (no smudges) and has clean lettering. Always keep a copy for yourself.

- Your article should be printed on $8\frac{1}{2}$- by 11-inch white bond paper. Margins should be at least 1 inch on all sides. Your name and address go at the top left corner of the title page; the word count, rounded to the nearest hundred, goes at the top right. The title, with your name beneath it, should be centered and start about one-third down the page. Your material should be double-spaced. Starting with page 2, your last name and a key word from the title should be placed at the top left. This is called the *page heading.* This information makes it easier for busy editors to find pages of your manuscript should they become separated. Number your pages at the bottom center or top right. Use paper clips to keep your manuscript together; never use staples.

- When you mail your manuscript, include a brief cover letter in which you introduce yourself.

- Remember that the competition is very stiff. Submit only your best work. If you receive a rejection, meaning your work is returned to you because the magazine cannot use it, do not give up. All writers suffer rejection. Keep trying.

To Be Submitted

A finished copy of your article.

Name _____

DATA SHEET 30.2

A Sample Query Letter

Author's Name
Address
City, State ZIP Code
Date

Editor's Name
Name of Magazine
Address
City, State ZIP Code

Dear [Editor's Name]:

Studying for math tests is a problem for many math students. Most just don't know how to study.

At Tom Paine High School, students have organized a math study club. The purpose of the club is to help students learn how to study for math tests and quizzes by introducing students to various study strategies. The club boasts a great success rate. All members have seen their grades improve, and almost half have seen their grades rise 15 percent or more. The club has recently begun to help interested groups organize study clubs for other subjects.

Would [name of magazine] be interested in an article describing how the Math Study Club at Tom Paine High School helps students improve their math grades? I would include interviews with club members to highlight the writing.

I feel I am very qualified to write this article. I am one of the founders of the club and serve as its current president.

Thank you for your time.

Sincerely,

Author's Name

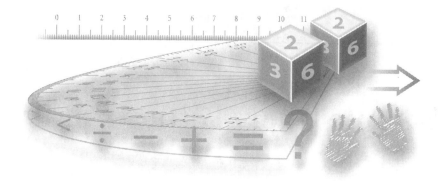

Keeping a math journal

A math journal contains written accounts of a student's ideas, questions, failures, successes, and musings regarding mathematics. It becomes a record of his or her experiences and progress. Journals are generally written in the first person and explore a variety of topics and issues: summaries of important concepts, frustrations, solutions to particular problems, or simply the student's impressions, opinions, or observations about math. Sometimes journal entries are rather general; sometimes they are personal. Always they present the student with the opportunity to explore his or her understanding of math.

Since journals can provide a written record of a student's growth in learning math, entries are often an excellent addition to a math portfolio. A math journal provides you with a means of getting to know your students in a unique way.

Goal

Each student will maintain a math journal for a marking period, semester, or school year. *Suggested time:* A partial class period to introduce the project and get students started.

Math Skills to Highlight

Writing as a means by which to express ideas about math. Specific math skills are dependent on the unit of study.

Special Materials/Equipment

A spiral notebook or composition book for each student to use as a journal.

Development

Before you begin this project, you must consider and decide on some guidelines for journals. How often should entries be made? How often will you look at the journals? Will students be encouraged to share entries with others? If students are maintaining portfolios, will they be encouraged to select journal entries for inclusion with their portfolios? Will there be any limits to the types of entries?

We recommend that you review student journals periodically. This gives you the opportunity to respond to students and make sure that they are writing in their journals regularly. You may collect journals every few weeks or get in the habit of reading a few from each class two or three times per week. This kind of schedule makes your workload more manageable.

We do not recommend grading or correcting the writing in journals. Grading will cause students to write what they think you want to see rather than take risks with their ideas and write about concepts or ideas they are not sure of. You should respond to the writing of your students by offering suggestions, constructive comments, questions, and encouragement. It is not unusual for students to write back to you in their journals, and you and they may even correspond through journals.

- Begin this project by explaining the purpose of a math journal. Distribute copies of Student Guide 31.1, and review the information with your students.
- Discuss how often students should make entries. While some students will write in their journals regularly without your prompting, others will need guidelines. An average of three times a week is a reasonable goal.
- Tell students how often you will collect journals.
- Mention that while you will not be grading journals, you will offer comments. Encourage students to respond to your comments. Tell them that while you will respect their ideas and privacy, you must report anything you read that you feel endangers the student or someone else.
- Emphasize that a math journal is a place to write about reflections and insights regarding math.

- Since many students may be unfamiliar with the types of entries that may be included in a journal, hand out copies of Data Sheet 31.2, "Some Ideas for Math Journals." Note that these are just a few examples of topics students might consider writing about.

Wrap-Up

Review journals periodically, and respond to the writing of your students.

Extension

Suggest that students select some of their best entries, polish and publish them in a math or class magazine or school newspaper, or include them in a math portfolio.

STUDENT GUIDE 31.1

Keeping a Math Journal

Situation/Problem

You will maintain a math journal in which you write about topics and issues in math.

Possible Strategies

1. Respond to math assignments, class work, and projects in your journal.
2. Select topics of interest in math that you would like to write about: things that are important to you; the solutions of specific problems; questions you may have; summaries of concepts, insights, and reflections; or your opinions about issues in math.

Special Considerations

- Use a standard spiral notebook or composition book for your journal. Be sure your name and class are on the cover. If you run out of pages, continue your journal in a similar book. Number your journals.
- Use your math journal only for math entries. Do not use it for other subjects.
- Bring your journal to class each day. Feel free to write in it at home as well as in school.
- Strive to write in your journal at least three times each week. Of course, you may write in it every day.
- Date your entries.
- Remember that your teacher will periodically read your journals and offer suggestions and comments.
- Share some of what you feel are your best entries with others.
- Review your journal from time to time, and reflect on your growth in mathematics.

To Be Submitted

The math journal.

Some Ideas for Math Journals

Following are some examples of topics you might wish to consider writing about in your math journal. There are countless others, limited only by your imagination and interest in math.

- Describe steps to solving a specific problem.

- Write about a concept or idea you find puzzling or fascinating. Explain why it interests you.

- Select a concept, and explain it.

- Describe a time when you used math outside school.

- Write about your greatest triumph in math.

- Describe how you used a computer to help you solve a math problem.

- Explain what makes a particular problem especially frustrating.

- Write on what you do not like about a topic in math.

- Explain situations when calculators are useful.

- Write about why estimation is an essential skill.

- Write about your feelings regarding a problem, situation, topic, or issue in math.

- Write about how you felt in math today.

- Write about shortcuts to solutions.

- Write about what you understand now that you had trouble with before.

- Write about your progress in math.

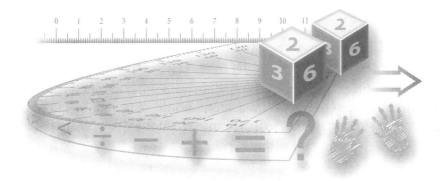

Math portfolios

A math portfolio is a collection of various kinds of a student's work gathered over a period of time. It may be compared to an artist's portfolio, which contains samples of the work an artist does, but in this case the portfolio holds examples of math work. A student's portfolio may include items such as assignments, quizzes, tests, reports, writing samples, projects, and project summaries. A good portfolio becomes a repository of a student's work, revealing not only the student's overall progress in math but also his or her attitudes toward mathematics. It can provide a far broader scope of a student's achievement than just tests and quizzes. Regular review of a portfolio can be helpful for students, teachers, parents, and administrators.

There are two types of portfolios: a *work portfolio* and an *assessment portfolio*. While some teachers require students to maintain only a general work portfolio, others require students to periodically select what they feel are their best papers from their work portfolios and compile an assessment portfolio for evaluation. Specific criteria are used in evaluation. (See Data Sheet 32.2, "Portfolio Assessment Criteria.") Whether you decide to have students maintain one or two portfolios, you undoubtedly will find that portfolios are an important tool in a math curriculum.

245

Goal

Each student will maintain a math work portfolio and an assessment portfolio for a specific length of time—perhaps a marking period, semester, or the whole year. *Suggested time:* A partial class period to introduce the project and get students started.

Math Skills to Highlight

Evidence of specific math skills will vary according to the unit of study. Portfolio items should reflect examples of problem-solving strategies that show evidence of student progress, growth, and reasoning.

Special Materials/Equipment

Two heavy envelopes or folders for each student: one to serve as a work portfolio and the other as an assessment portfolio; a milk crate or box for each class to store the portfolios; a felt-tipped pen to write students' names on portfolios.

Development

Decide whether you will require students to maintain only a work portfolio or a work portfolio and an assessment portfolio. We suggest both. Since not all of the items that go into a work portfolio will be examples of a student's best efforts, an assessment portfolio offers students a chance to select examples of their work that show their greatest progress and learning. You must also decide what type of work students will file in their portfolios and how long they will maintain their portfolios.

- Start this project by explaining the purpose of a portfolio to your students. You may find that some of your students have never been required to maintain one before and will be uncertain of what they are to do.

- Distribute copies of Student Guide 32.1, and review the information with your students. Especially note the types of material that may be placed in a portfolio.

- Hand out copies of Data Sheet 32.2, "Portfolio Assessment Criteria." Explain that the abilities and skills noted on the sheet are the criteria you will concentrate on when you review the portfolios. (Some teachers prefer to focus their assessment efforts on two or three specific criteria. They feel this helps them to remain more objective.) Of course, you may include additional criteria based on your program and needs.

- Distribute large envelopes or folders to serve as the portfolios. Have students clearly print their names on their portfolios with a felt-tipped pen. If possible, provide a milk crate or box in the classroom to store the portfolios. Keep the portfolios in alphabetical order, and instruct students to file their own work. This relieves you of the burden of filing. If you cannot store portfolios in the classroom or another central location, you may permit students to maintain them at home.

- Emphasize that any work put into the portfolio should be dated and placed in sequential order according to the date.

- At the end of the specific time frame (a unit or marking period, for example), have students review their work and select items to place in an assessment portfolio. You may provide guidelines as to the number of items to include. You may also require that specific assignments be included so there is some consistency among the portfolios.

Wrap-Up

Review portfolios with individual students, and discuss their growth in math. Use the assessment portfolios as a means to focus on future achievement.

A note regarding grading: Since most items that go into a portfolio are already graded, it is not usually advisable to place a grade on the portfolio. A portfolio's greatest value lies in identifying student growth and mastery of mathematical concepts and skills. Moreover, when students know that a portfolio will be graded, many will not choose work that shows growth or reveals original thinking, but instead will select samples that they feel are most likely to result in the best grade.

Extensions

If you had students maintain portfolios for only a unit or marking period, consider extending the time frame. Share portfolios with parents and administrators.

STUDENT GUIDE 32.1
Math Portfolios

Situation/Problem

You will create and maintain a math portfolio. Your portfolio will show your progress in math through various samples of your work.

Possible Strategy

Select examples of your work in math to be stored in your portfolio.

Math Portfolios *(Cont'd.)*

Special Considerations

- While you may select work for your portfolio, check with your teacher to find out if he or she requires that some specific examples of your work be included. Some of the work you may wish to put in your portfolio follows:

  Solutions to open-ended questions

  A report

  A math project

  A summary of your contribution to a math project

  An article you have written about a topic in math

  Homework

  Tests

  Quizzes

  Work that was done in another class but relates to math

  A math problem you have written

  An explanation of a math concept

  An entry from your math journal (if you are keeping a journal)

  Comments from teachers

  Any work required by your teacher

- Be sure to place all required math work in your portfolio.
- Make sure your name is on your work, and date all assignments.
- File all assignments in chronological order.
- Clip or staple multipage assignments together so that individual sheets do not become separated or lost.
- Include a Contents page, which lists the items in your portfolio.
- Include a letter of introduction to the reader of your portfolio. Your "reader" may be your teacher, your parent, or an administrator. This letter should contain:

  An explanation of what you chose to put in your portfolio and why you chose these items

  A description of the major concepts your portfolio illustrates

  How this portfolio shows your progress

  What work you liked best, and why

To Be Submitted

A portfolio, including a table of contents and letter of introduction.

DATA SHEET 32.2

Portfolio Assessment Criteria

When I review your portfolio, I will be looking at the following criteria to show me your overall growth in learning math:

- Understanding specific problems in math
- Understanding specific concepts
- Evidence of effective mathematical reasoning
- The ability to choose effective problem-solving strategies
- The ability to gather, analyze, and organize data in the solving of problems
- The ability to interpret results and draw conclusions
- The ability to support conclusions
- The use of appropriate math vocabulary, notation, and labels in written work
- The construction, manipulation, and understanding of models
- The use of technology in problem solving
- Evidence of accuracy in applications and computation
- Evidence of self-assessment of work
- Evidence of critical thinking
- Enthusiasm in the learning of mathematics
- Understanding and appreciation for the wide scope of mathematics in our world

Math and Art and Music

Making a math poster

Math posters can make a classroom more attractive and interesting. This is especially true when students design and make posters that focus on a math idea, concept, formula, or term.

Goal

Working individually or in pairs, students will create a math poster for the classroom or school. The subject matter is up to them. *Suggested time:* One to two class periods.

Math Skills to Highlight

1. Representing mathematics in the format of a poster
2. Using math as a means to communicate ideas
3. Using technology in problem solving

Special Materials/Equipment

Poster paper; rulers; colored pencils; markers; sample posters. Old magazines, catalogues, and clip-art books can be good sources of illustrations for posters. *Optional:* T-squares; drafting materials; computers and printers for writing and printing text; Internet access for research.

Development

Show students a variety of posters. Math posters would be most useful, although any posters will be helpful. Explain that the purpose of a poster is to share an idea. Effective posters are visually catchy. They gain attention by attracting the reader's eye with vivid colors, interesting photos or pictures, or compelling headlines. If you are showing examples of posters, ask students to point out what they find most interesting about each poster. Ask what they feel the poster is trying to convey.

- Begin the project by telling students that they will work alone or with a partner to create a math poster.
- Distribute copies of Student Guide 33.1, and review the information with your students. Especially note the possible topics for posters. Ask students to suggest more possible topics, and list them on the board or on an overhead projector. These examples will help students get started.
- Hand out copies of Data Sheet 33.2, "Pointers About Posters." Review the material with your students, and discuss the suggestions for creating effective posters. Emphasize that their posters should contain as many of the components of effective posters as possible.
- Discuss the importance of neatness and clarity. A poster's message might be negated if the poster is sloppy, or the message of a neat poster might be lost if it is bundled with too much information.
- Encourage students to design their posters in a way that best shares their ideas about math.

Wrap-Up

Display the finished posters in your classroom. If there is not enough space, perhaps a hallway or section of the library might be used.

Extension

You may want to have a school or class vote on the best poster. Award math prizes for the first-, second-, and third-place winners.

STUDENT GUIDE 33.1

Making a Math Poster

Situation/Problem

You are to design a math poster to be displayed in your math class or elsewhere in your school.

Possible Strategies

1. Study examples of posters. Determine what you think makes some posters more effective than others.
2. Review Data Sheet 33.2, and compare what you think makes an effective poster to the information on the sheet.
3. Browse through your math text, consult other reference books, or check online sources to find an idea, concept, or formula that you can convey in a poster.

255

Making a Math Poster *(Cont'd.)*

Special Considerations

- Be sure the topic that you select is not too broad or complex to be communicated in a poster.
- If your topic at first seems too broad, try narrowing it down. Perhaps you can pick part of it to illustrate.
- Following are some samples of possible topics for posters. There are many more:

    Properties in Math

    Exponents

    Simplifying Fractions

    Decimals

    Percentages

    Scientific Notation

    Rounding

    Area Formulas

    Circumference

    Perimeter

    Types of Triangles

    Types of Polygons

    Types of Quadrilaterals

    Parts of Circles

    Types of Lines

    Math and Your Future

    The Importance of Math

- Be sure to review Data Sheet 33.2. It will help you create an effective poster.
- Try to make your poster unified and balanced.
- Try to create a clever illustration. Consider hand-drawn illustrations or clip art. Some computer programs offer clip art you might use.
- Be neat. Make sure your illustrations, graphics, and lettering are precise and arranged attractively. Lines should be straight.
- If you have drafting tools, you may wish to use them.
- If you enjoy calligraphy, you may wish to use this type of lettering to highlight your poster.
- Write text on a computer, and print it in an appealing font.

To Be Submitted

Your poster.

DATA SHEET 33.2

Pointers About Posters

We see posters everywhere, from a store window advertising an upcoming local event to movie posters announcing the coming attractions. A poster is usually larger than a standard sheet of writing paper and usually, but not always, includes a picture or graphic design with the text. The elements of effective posters are listed below. You should try to incorporate them in your poster.

- *Selling point.* The purpose of any poster is to announce something or sell something.

   The message may be direct and include words such as buy, ask, or attend.

   The message may simply announce or communicate an important idea.

- *Benefits.* The poster tells readers what they have to gain by following the poster's advice or suggestions, or at least offers important information.

- *Gaining attention.* Every poster attempts to gain and hold attention. It may do this in several ways:

   The poster may employ bold headlines or amusing or eye-catching pictures.

   Posters may use dramatic situations, slogans, a play on words, rhymes, or popular sayings, or they may appeal to the conscience of people.

   The element of surprise helps to gain and keep attention. This may include unexpected, bizarre, or exaggerated illustrations, or pictures taken from unusual perspectives or dramatic angles, such as tilting the picture on its edge.

- *Simplicity.* The best posters are simple and easy to read. Too many colors, pictures, or fancy lettering detract from the poster's message.

- *Unity.* Everything in a poster should work toward its purpose.

   Main ideas may be highlighted by using devices such as arrows, dots, or pointing fingers.

   One or several elements in the layout may touch or overlay others.

   The poster may be surrounded with a border.

   The background may be painted, or the poster may be on colored paper.

- *Balance.* No part of a poster should overpower another. Balance can appear in two ways: formal balance and informal balance.

   Formal balance: A line of symmetry divides the poster, where one half balances the other.

   Informal balance: The various elements of the poster, such as size, color, and shape, are in harmony and give an impression of being in balance.

- *Workmanship.* This is the overall quality of the poster, including lettering, coloring, pictures, and message. Quality workmanship results in outstanding posters.

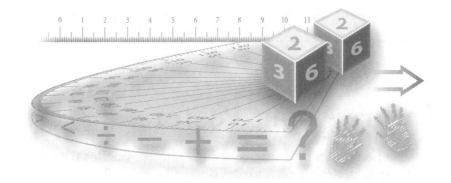

Creating a logo

A logo is a visual representation that establishes an identity. Since effective cooperative learning depends on a feeling of interconnectedness, this project can be an excellent team-building activity.

Goal

Working in groups of three to five, students will think about their group and design a logo that best portrays or describes them. Students will show their logo to the class, and explain why they chose their design. *Suggested time:* Two to three class periods.

Math Skills to Highlight

1. Making and testing conjectures
2. Communicating math strengths and goals
3. Synthesizing information related to math
4. Using technology in problem solving

Special Materials/Equipment

Colored pencils; markers; compasses; rulers; stencils; $8\frac{1}{2}$- by 11-inch white paper; sample logos. *Optional:* Computers and printers; Internet access for research.

Development

Since students will create logos that will represent their group, this project works best after students have worked together previously as a group. Students will be better able to identify their individual and collective strengths and weaknesses. Prior to starting this project, cut out logos from advertisements and product packages and bring them to class.

- Introduce this project by explaining to your students that they will work in groups and create a logo for their group.

- Discuss logos as visual representations of organizations, companies, or products. Logos convey a message and remind people of the group or product they represent. Show students examples of logos and ask them if they can identify the company they represent. In many cases, they will be able to. This demonstrates the power of logos. (You may also direct students to the Web sites of major companies and organizations to view logos.)

- Hand out copies of Student Guide 34.1, and review the information with your students.

- Distribute copies of Data Sheet 34.2, "Creating an Effective Logo," and discuss the material on the sheet. Emphasize that students should refer to this sheet when creating their logos.

- Provide students with the time necessary to discuss the characteristics of their group, which they may incorporate in their logo. Stress that each student should contribute to the discussion. Consider requiring each student to contribute his or her own design. The group then selects the best design, or they may wish to combine the best features of two or more designs. However they pick their logo, students should have reasons for their choice.

- Suggest that students design their logos on computers. Many art and drawing programs have the capability to produce outstanding logos.

- Note that the logos may also be drawn and illustrated by hand. Encourage students to color their logos.

- Remind students that they should choose a spokesperson to explain the selection of their logo to the class.

Creating a logo

Wrap-Up

Each group introduces its logo and explains its design. You may wish to display the logos in the classroom or elsewhere in your school.

Extension

Use a silk-screening technique to print each group's logo on T-shirts. Your school's art teacher may be willing to help you with this.

STUDENT GUIDE 34.1

Creating a Logo

Situation/Problem

You and your group are to design a logo to represent your group. This logo will be drawn on $8\frac{1}{2}$- by 11-inch paper and will project an image of your group. After finishing your logo, you will show it to the class and explain why you chose its design.

Possible Strategies

1. Study various logos. Discuss what makes some more effective than others.
2. Brainstorm with your group to determine your unique characteristics as both individuals and a group.
3. Review Data Sheet 34.2 to learn about the elements of effective logos.
4. Individually create rough sketches of what each of you thinks your group logo should be like. Discuss the logos and pick the best one, or combine features of several.

Special Considerations

- As you study various logos, try to decide what message is communicated, what you liked about it, what makes it memorable, and what you do not like about it.
- To find the characteristics of your group, ask yourself questions like the following:

    What are our strengths?

    What are our goals?

    How do we differ from other groups?

    How do we differ from each other?

    What do we wish to share or communicate through our logo?

Creating a Logo *(Cont'd.)*

- Give your group a name.
- Design a logo that best describes your group. Sketch several possible logos. Compare them, and select the best one.
- When creating your logo, pay attention to design, especially the spacing between letters and pictures. Your logo should be visually appealing.
- Write down the reasons that you selected the logo you did. Select a spokesperson who will later share your reasons with the class.
- You may wish to design your logo on a computer. Many art and drawing programs have the capability to create logos.
- You may also hand-draw your logo on white paper. Be neat with your artwork and lettering.

To Be Submitted

Your logo.

Notes

Name _____

Creating an Effective Logo

A logo is a visual representation that conveys a meaning. It may be a picture, a name, or a picture and a name. You probably would recognize many, including the NBC Peacock, MTV graphic, the Hallmark Crown, and the Golden Arches of McDonald's. Big companies pay hundreds of thousands of dollars to consultants to design logos that will provide positive, memorable impressions. Following is some information that will help you design your own logo.

Qualities of an Effective Logo

1. Makes a good first impression
2. Represents who you are and your ideas and attitudes
3. Possesses something unique or interesting to help you stand out from the crowd—a mark of distinction

How to Design an Effective Logo

1. Think of a symbol that uniquely depicts you. This symbol may be an illustration, a name, or a combination of both.
2. Carefully consider the lettering you will use—for example:

    Italic type (*slanted*) denotes action or speed and projects a modern or progressive image.

    Capital letters suggest formality and steadiness.

    Lowercase letters suggest an informal manner or casual image.

    Outlined letters project an informal image.

    Thin letters denote professionalism.

    Thick or bold letters project strength or dependability.

    Script denotes gentleness or caring.

3. Keep your logo clear and simple. Avoid cluttering it with unnecessary letters or art. Eliminate any unnecessary details.
4. Strive for balance in your spacing of letters, pictures, and designs.
5. Be neat in drawing and sketching.

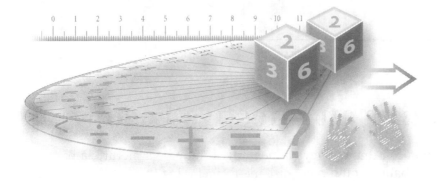

I wanna be like Escher

Maurits Cornelis Escher (1898–1972) was a Dutch graphic artist. Although his early studies focused on architecture, he soon turned to graphics. Until 1937 he drew and sketched mostly landscapes, but then concentrated on constructing images that exist only on paper or in abstract theory.

While his graphics may be grouped according to several different themes, the division of the plane is the focus of this project. A tiling design that covers a plane with no gaps or overlaps is called a *tessellation*. A *pure tessellation* is a design in which only one figure is used. A *regular tessellation* uses only one regular polygon to tile the plane. A *semiregular tessellation* is a design that covers the plane using two or more regular polygons. Escher created over 150 different tessellations in his drawings, yet he had no formal mathematical training.

Goal

Working individually, students will make a drawing similar to those Escher used to tile a plane. *Suggested time:* Three to four class periods.

Math Skills to Highlight

1. Defining and discussing tiling the plane, tessellations, including pure tessellations, regular tessellations, and semiregular tessellations.

2. Reviewing the names of polygons, including triangle, quadrilateral, rectangle, square, parallelogram, rhombus, and regular polygon.

3. Using the protractor to measure angles, or using the formula $[180(n-2)]/n$ to find the measure of each interior angle of a regular polygon. n stands for the number of sides.

4. Discussing transformations, including slides, reflections, slide reflections, and rotations.

5. Using technology in problem solving.

Special Materials/Equipment

Books containing Escher's graphics; protractors; scissors; envelopes to store patterns; rulers; transparent tape. *Optional:* Colored pencils or thin markers; computers and printers; Internet access for research.

Development

Show students examples of Escher's graphics in books about Escher. An excellent source is *Fantasy and Symmetry: The Periodic Drawings of M. C. Escher* by Caroline H. MacGillavry (New York: Abrams, 1976). Another is *M. C. Escher: His Life and Work* by J. L. Locher (New York: Abrams, 1992). Many other sources are probably in your local library. Numerous Web sites devoted to Escher and his work may be found by conducting a simple search using "M. C. Escher" or "tessellations."

- Begin the project by discussing pure, regular, and semiregular tessellations and the differences among them.

- Distribute copies of Student Guide 35.1, and review the information with your students. Note that the guide provides step-by-step instructions for students.

- Hand out copies of Data Sheet 35.2, "Polygon Patterns." Ask students to cut out the large regular triangle, square, and other regular polygons. Distribute an envelope to each student, and instruct students to keep all of their pieces in it to prevent the pieces from becoming lost. The small squares and triangles (the ones that have a 1-inch side) will be used only in the extension for this project. Do not cut them out yet.

- You may permit students to work in small groups to determine which regular polygons will make a regular tessellation. Remind students that only one shape may be used per design. Polygons may not overlap, and no spaces should be left between the tiles. Students should discover that only the regular triangle, square, and regular hexagon tile.

- Explore why some figures tile. The key here is the measurement of the angles. You may ask students to measure the angles of the regular polygons, or you may use the formula $[180(n - 2)]/n$ to determine the measure of an interior angle of a regular polygon. Students should discover that in order for a figure to tile, the sum of the angles around any point is 360 degrees. As an option, students may generalize that any triangle and quadrilateral will tile. They may require slides, reflections, slide reflections, or rotations.

- Distribute copies of Data Sheet 35.3, "Steps to Making an Escher-Like Drawing." These steps are illustrated with examples for clarity. Be sure your students understand the steps.

- Hand out copies of Worksheet 35.4, "Isometric Dot Paper." Students are to cover the worksheet by sliding, reflecting, sliding and reflecting, or rotating the shape. With details and color added, the drawings should resemble some of Escher's works.

Wrap-Up

Students should share their drawings with other members of the class. You may wish to display drawings in a central location.

Extensions

By using the smaller regular triangle, square, and other regular polygons, students can create semiregular tessellations through manipulating the patterns. Combinations may include:

- Octagon and square
- Hexagon, square, and triangle
- Hexagon and triangle
- Square and triangle

Students may also wish to make an Escher-like drawing, consisting of two figures.

266

STUDENT GUIDE 35.1

I Wanna Be Like Escher

Situation/Problem

Maurits Cornelis Escher (1898–1972) was a graphic artist. One of his most famous themes was tiling the plane. A tiling design that covers the plane with no gaps or overlaps is called a *tessellation*. Escher created over 150 different tessellations in his drawings. His designs are truly fascinating. In this project, you will create a tessellation of your own. When your drawing is complete, share it with your class.

Possible Strategies

1. Study examples of Escher's graphics on tiling in both print and online sources.
2. Determine which polygons will make a regular tessellation. List them.
3. Design a shape of your own by using a polygon that will tessellate.

Special Considerations

- From Data Sheet 35.2, choose a polygon that will form a regular tessellation. You may select a triangle, square, or hexagon.
- Refer to Data Sheet 35.3 for instructions on how to make an Escher-like drawing. Follow the suggestions carefully.
- Use Worksheet 35.4 for your drawing. Be creative.
- You may color your design.

To Be Submitted

Your worksheet.

Name _____

Polygon Patterns

Regular Triangles

Squares

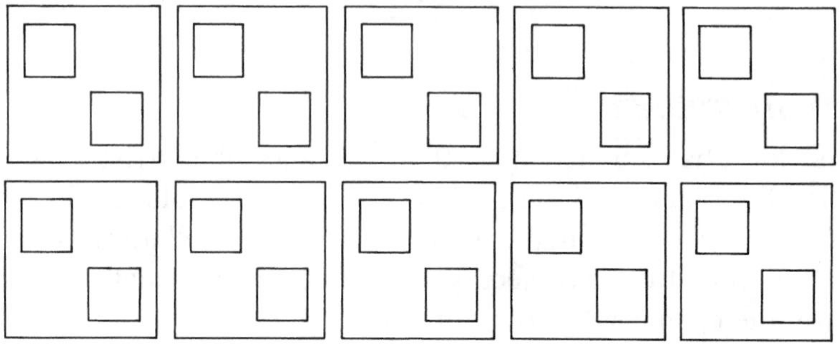

Regular Pentagons

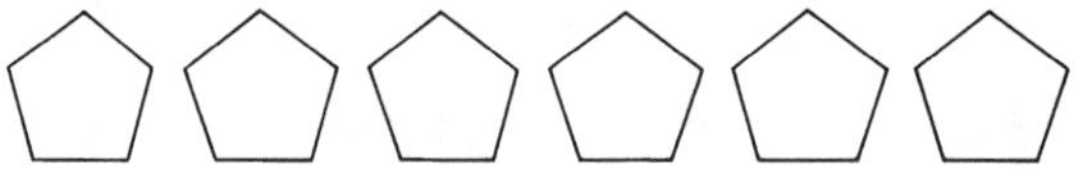

Regular Hexagons

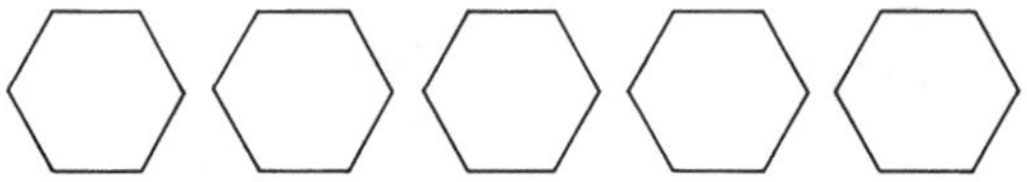

Regular Octagons

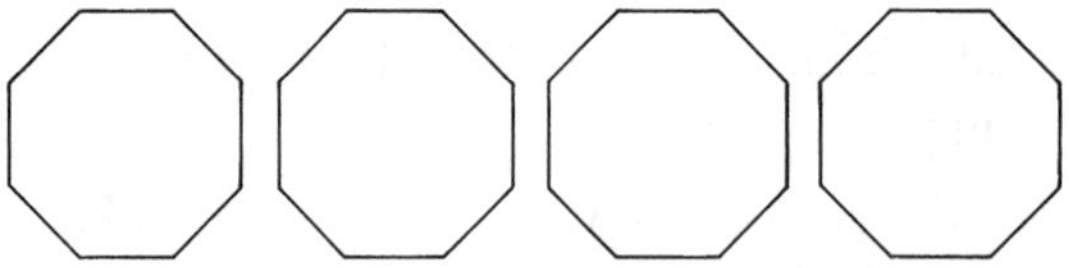

Steps to Making an Escher-Like Drawing

1. Start with a shape that will tessellate. A hexagon is used in the example.

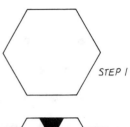

STEP 1

2. Cut a portion of the figure on one side, slide it to the opposite side, and tape it. In the example, this was done three times.

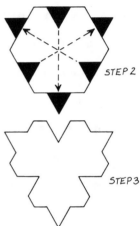

STEP 2

3. What does the shape resemble? In this case, the shape resembles a snowflake.

STEP 3

4. Cover the paper by sliding, reflecting, sliding and reflecting, and/or rotating the shape.

5. Add details.

STEPS 4 and 5

269

WORKSHEET 35.4

Isometric Dot Paper

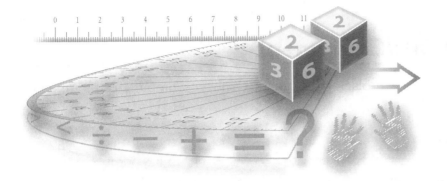

The Plus and Minus comic strip

Comic strips are regular features in most newspapers. Combining unique characters, witty dialogue, and good art, popular comic strips attract a strong following and transcend generations of readers. This project gives your students a chance to create comic strips with a focus on math.

Goal

Students will work in pairs or groups of three to make a math comic strip. *Suggested time:* Two to three class periods.

Math Skills to Highlight

1. An awareness of spatial relationships.
2. Using technology in problem solving.
3. Other skills will vary, depending on the story line students select for their comic strip.

Special Materials/Equipment

White drawing paper; black pens; rulers. *Optional:* Computers and printers.

Development

Consider collaborating with your students' art teacher for this project. He or she might permit students to work on their comic strips in art class. If you arrange such collaboration, you should provide the introduction to the project and help students focus their story line on mathematics, while the art teacher manages the actual creation of the comic strips.

Before beginning the project, collect examples of comics from newspapers and magazines. Since some math texts intersperse comics and cartoons throughout their chapters, go through your text and list pages with cartoons or comics. Providing students with examples of comics that have a mathematics slant will help them generate ideas for their own comic strips.

- Start this project by explaining that students are to work in pairs or small groups and create a comic strip focusing on some aspect or idea in mathematics. Encourage students to invent their own characters; however, if they wish, they may use the characters Plus and Minus in their comic strips.

- If you have examples of comic strips, distribute them. Point out how comic strips usually focus on an idea that is developed in a narrative and usually ends with a punch line. Also note the artwork. Cartoonists must remain aware of spatial relationships as they work. Because the action must fit within a box, the artwork must be carefully designed. Too much detail crowds the box; too little may not be visually appealing or may not express the cartoonist's ideas.

- Hand out copies of Student Guide 36.1, and review the information with your students. Especially note the points in the Special Considerations section, which help students to generate ideas for their comic strips. Emphasize that the comic strips must have a math angle or punch line.

- Recommend that students keep their comic strips to four frames or fewer. More than that will require a rather complex narrative and artwork.

- Some students may protest that they cannot draw well. Tell them that you do not expect them to be professional artists. Their ideas are more important than the expertise of the drawing.

- Distribute copies of Data Sheet 36.2, "Comic Tips." Discuss the information with your students, which offers background on some of the techniques cartoonists use.

- Encourage students to create their comic strips on computers, especially if they have access to art and drawing software.

Wrap-Up

Display the comic strips in class, or make copies and bind them in a class comic book. Hand out copies of the book to students, and be sure to display some in your school's media center.

Extension

Some students may wish to explore creating flip books. A good source is *Making "Movies" Without a Camera* by Lafe Locke (Cincinnati, Ohio: Betterway Books, 1992). Another is *Copier Creations* by Paul Fleischman (New York: HarperCollins, 1993). Your local library probably has several more.

STUDENT GUIDE 36.1

The Plus and Minus Comic Strip

Situation/Problem

You and your partner or group are to create a comic strip that focuses on math or uses a math idea or concept in some way. You may use Plus and Minus as your main characters, or you may create your own.

Possible Strategies

1. Study various comic strips. Identify their features, and analyze how they are done.
2. Consider inventing your own characters. Brainstorm possible characters for your comic strip.

Special Considerations

- Identify an idea in math that can be used in a comic strip.
- Think about your characters, what they are like, and how they relate to math. What situations could you create that show them using or discussing math?
- Brainstorm ideas. Jot down ideas, possible topics, phrases, and words. Doodle to generate ideas. Use your imagination. Let your thoughts roam. Think of associations. Look for a play on words or funny circumstances. Try to make connections.
- Look through your math text for ideas.
- Once you have an idea, sketch it out on paper. Try keeping your comic strip to four frames or fewer. Be aware of spatial relationships. Since you are limited by the size of each frame, you must design your art carefully. Try to keep things in proportion.
- Remember that you will need room for the narrative and dialogue.
- Refer to Data Sheet 36.2 for suggestions for creating a comic strip.

The Plus and Minus Comic Strip *(Cont'd.)*

- You may need to try several sketches before you are pleased with your strip.
- When drawing, use rulers to make straight lines, and be neat with your artwork. Write the dialogue clearly.
- If your computer has art and drawing software, you may wish to create your comic strip on your computer.

To Be Submitted

Your comic strip.

Notes

DATA SHEET 36.2

Comic Tips

From *Peanuts* to *Dick Tracy, The Wizard of Id* to *Doonesbury,* comics vary. Yet all have common elements. The most important include:

- **A narrative related by a series of pictures.** Called *frames* or *panels,* the narrative is usually humorous but may be serious.
- **Continuing characters.** The characters are unique and have their own personalities. They may be human, animal, or even aliens.
- **Dialogue** within the pictures, usually in the form of "speech balloons" from the characters' mouths.
- **Captions** that tell the story. This is called the *narrative.*
- **Background art** that enhances the scene but does not overshadow the characters.
- **A punch line:** a joke, an amusing situation, an insight or observation, or a play on words.

Creating a Comic Strip

1. Develop a cast: people, animals, or other characters.
2. Write dialogue in the form of speech balloons. You may use captions to describe the action. (Not all comics have dialogue or captions. These rely on the art to deliver the punch line.)
3. Provide a setting, that is, a specific place where the story occurs.
4. Add a narrative that shows action in proper sequence.
5. End with a punch line developed out of the narrative.
6. Be selective in deciding what to put in and what to leave out of your comic strip. Remember that you have a limited amount of space.

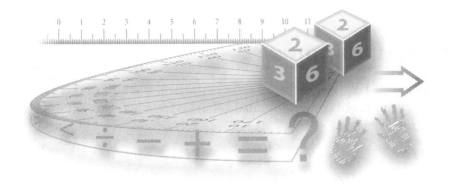

Numbers and songs

A novel and enjoyable way for students to highlight math concepts or facts is to write a song about them. Even if they do not have any particular musical talent, most students plunge into this project with great enthusiasm.

Goal

Working in groups of two to four, students will compose songs that express an idea in mathematics. On completion of the project, students will be encouraged to perform or record their song for the class. *Suggested time:* Two to three class periods.

Math Skills to Highlight

Skills will vary depending on student songs.

Special Materials/Equipment

Rhyming dictionaries. For students who have a musical background, you may wish to provide examples of songs and music sheets. *Optional:* Computers with Internet access for research.

Development

Consider working with your students' music teacher for this project. While you work with students on the math for this project, he or she can help them write songs.

- Begin this project by explaining to your students that they will work in groups to write a song that identifies a math concept or fact. If you have students who study music or play musical instruments, try to organize groups so that each group has at least one of these students. Students who have a musical background can assume roles of leadership in this project.

- Distribute copies of Student Guide 37.1, and review the information with your students. Encourage them to select the type of music they like, and write their song in that style. Writing a song will be easier if they choose a type of music with which they are familiar.

- Hand out copies of Data Sheet 37.2, "Tips for Writing Lyrics and Music," and discuss the information with the class. Note that students who do not have a musical background might simply compose a song using a beat, much like a rap song.

- Discuss that most songs use lyrics that rhyme. If you have access to rhyming dictionaries, encourage students to use them. If these dictionaries are not available, encourage students to generate their own lists of rhyming words. Using the board or an overhead projector, ask students to help you create examples of rhyming words, such as the ones below:

  run, fun, sun, son, done, none, ton

  high, sigh, cry, tie, pie, lie, try, multiply

  dad, pad, tad, bad, add, sad, fad

  hide, cried, tried, divide, wide, sighed

  The lists can go on and on. Do a few examples, and your students will understand how to continue.

- You might consider suggesting that students visit Web sites devoted to music. Many sites can provide excellent background and information for composing a song, as well as sample songs. *Caution:* Make sure that students do not download songs from sites that require fees or violate copyrights. Also, make sure that students do not simply listen to songs instead of completing the project.

- For students who have a musical background and want to set their lyrics to musical notes, hand out copies of Worksheet 37.3, "Music Sheet." Suggest that if they play musical instruments, they might use their instruments to match the notes to their lyrics. Note, however, that only students who have the interest and musical ability will be able to manage this.

- If groups have a member who plays a musical instrument, encourage the group to include the use of the instrument with their song.

- If some groups have no members who play an instrument, assure them that using only lyrics in their songs is fine.

- If your students have access to a computer and software that allows song-writing, encourage them to use it. It is truly astonishing what students can do using such equipment. (Your students' music teacher may have this type of equipment.)

- Suggest that students videotape or tape-record their songs. Perhaps they can do this in the music room or at home. Be sure to follow the policies of your school regarding the videotaping of students.

Wrap-Up

Encourage students to perform their songs for the class. If a group is reluctant to perform, accept a tape or just a written copy of their song.

Extension

Videotape your students performing their songs in a Festival of Math Music. (Again, be sure to follow the policies of your school regarding videotaping students.)

STUDENT GUIDE 37.1

Numbers and Songs

Situation/Problem

You and your group are to compose and perform a song that highlights a concept or fact in mathematics.

Possible Strategies

1. Discuss favorite types of music with the members of your group. Select one that most group members enjoy. It will be easier to write a song in a familiar style of music.
2. Brainstorm math concepts or ideas. Have a recorder write down the group's ideas. Use your math text to generate ideas.

Special Considerations

- Choose a math concept or fact that can be expressed in a song.
- Refer to Data Sheet 37.2 for suggestions on songwriting.
- Remember that your song does not have to be long, but it should contain an idea about math.
- Use a rhyming dictionary to help you write lyrics that rhyme, or generate your own list of rhyming words.
- Consult Web sites that offer information about songwriting. Searching with the term "song writing" will result in a list of helpful sites. Do not download any songs from sites that require fees, and do not violate copyrights.
- If possible, write musical notes as well as lyrics for your song. This will be easier if a group member has had training in music or plays an instrument.

Numbers and Songs *(Cont'd.)*

- You may find it easier to write your lyrics according to a specific beat. Each syllable of a word in the lyrics would equal one beat. You may wish to try various beats.
- If a member or members of your group play a musical instrument, try using the instrument with your song.
- If you have access to a computer and software that support songwriting, try using them to compose your song.
- If necessary, meet outside class to rehearse performing your song.
- Tape-record or videotape your song.
- Consider performing your song live for the class.

To Be Submitted

A copy of your song.

Notes

_____

_____

_____

_____

_____

_____

_____

_____

_____

_____

_____

_____

_____

DATA SHEET 37.2

Tips for Writing Lyrics and Music

Songs are a form of communication in which a songwriter expresses ideas about a topic. Here are some suggestions for writing a song.

- The words of songs are called *lyrics*. Lyrics are similar to poems, but lyrics are set to music.
- Lyrics usually include rhyme. This helps make the song pleasing to the ear of the listener.
- A rhyming dictionary can be helpful in writing lyrics.
- Many popular songs contain a *chorus*—the part of the song that is repeated.
- *Rhythm* is the pattern of musical notes in a song.
- The *beat* is the underlying pulse in a piece of music. It may be fast or slow. The speed is called the *tempo*.

To write a song, try the following:

1. Find a topic or idea. It should be one that listeners can easily understand.
2. Compose lyrics that express your idea. Although many professional songwriters compose the music first, it is usually easier for beginners to write lyrics and then explore rhythm and beat.
3. Set your lyrics to a beat. The beat in a song is unchanging. Often the lyrics of a song fit a particular beat. Experiment with several possibilities. A four-beat count is one of the easiest to work with.
4. If you have an understanding of musical notes, write down the notes that go with the lyrics.
5. Create a videotape or tape a recording of the performance of your song.

Music Sheet

Write the lyrics and music for your song below.

Song Title: _____

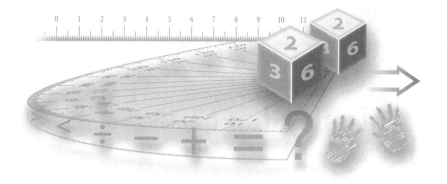

The math in music

When most people think of music, a favorite song, a rhythm, or the feelings a memorable melody evokes come to mind. Few think of mathematics, even though math is an integral part of music. This project gives students the opportunity to learn how music is based on numbers.

Goal

Working individually, students will research the ways mathematics is fundamental to music. They will then write an essay based on their research. *Suggested time:* One to two class periods, with additional research time outside class.

Math Skills to Highlight

1. Researching information about mathematics
2. Using writing to express ideas about math
3. Using technology in problem solving

Special Materials/Equipment

Reference books that contain information about music and math. *Optional:* Computers and printers; Internet access for research.

Development

This project will require some real digging for facts, and you may wish to assign it as a challenge.

- Start this project by explaining to your students that music is based on mathematics. An *interval*, for example, is the space between two notes. An *octave* is the space of eight notes or tones of a major or minor scale. The sound of two notes an octave apart is explained as the *frequency* of the higher note being twice that of the lower note. *Rhythm* is founded on the lengths of notes and the interrelationship between them. *Beats* are a metrical structure that provide the pulse of a piece of music. Some common beats in popular music are meters in $\frac{4}{4}$, $\frac{2}{4}$, $\frac{3}{4}$, and $\frac{6}{8}$ time.
- Distribute copies of Student Guide 38.1, and review the information with your students.
- Prior to beginning the project, consult your school librarian about this topic, and ask her to set aside reference books for your students. Schedule at least one period in the school library to help students get started with their research. Suggest that students continue their research outside class, using both print and online sources.
- Since students will likely require time outside class to complete this project, set a deadline. A week to ten days should be enough.
- Distribute copies of Data Sheet 38.2, "Tips for Essay Writing," and review the suggestions with your class.
- Encourage students to write their essays on computers.

Wrap-Up

Display the essays of your students.

Extension

Produce a book containing your students' essays. You may title it *Math in Music: A Collection of Student Essays.*

STUDENT GUIDE 38.1

The Math in Music

Situation/Problem

You are to research the role mathematics plays in music and write an essay based on the information you found.

Possible Strategies

1. Consult reference books about music, as well as online sources, and research the relationship between music and mathematics.
2. Focus your research on key words in music, including *octave, meter, beat, interval,* and *rhythm.*

Special Considerations

- Take accurate notes.
- Be sure to record your sources.
- Refer to Data Sheet 38.2 for suggestions on organizing and writing your essay.
- Write your essay on a computer, which will make revision easier.

To Be Submitted

Your finished essay.

DATA SHEET 38.2

Tips for Essay Writing

An *essay* is usually a short piece of writing in which the author focuses his or her efforts on a specific topic. Following are some guidelines for writing an effective essay:

- The typical essay uses a simple format:

    An *introduction,* in which the main point of the essay is stated.

    A *body,* which explains the main ideas. Depending on the length of the essay, the body might be a few paragraphs or several pages.

    A *conclusion,* which summarizes the main points of the essay.

- An essay should be written in a clear, concise style.

- In organizing your essay, identify the main ideas you found in your research. List them on a piece of paper. These main ideas will probably make up the body of your essay.

- Organize your main ideas in order, from most important to least important. This will be the basic outline of your essay.

- Support your main ideas with details and examples.

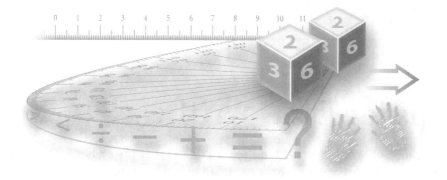

Making three-dimensional octahedra and classroom decorations

This project can add some pizzazz to your classroom, as well as your teaching. Students will likely enjoy the hands-on element of constructing three-dimensional octahedra that may serve as classroom decorations. The project is also a good introduction to solid geometry and the Platonic solids.

Goal

Students will work individually to construct classroom decorations. *Suggested time:* Two class periods.

Math Skills to Highlight

1. Inscribing an equilateral triangle in a circle
2. Identifying the parts of a circle
3. Identifying regular octahedra and their faces, edges, and vertices

4. Evaluating Euler's formula

5. Using technology in problem solving

Special Materials/Equipment

At least eight pictures per decoration, which relate to a theme; compasses; straightedges; scissors; glue; big paper clips or binder clips; a hole puncher; poster paper or light cardboard; cord for hanging the decorations. Old magazines, catalogues, brochures, newspapers, and even greeting cards and wrapping paper are good sources for pictures. *Optional:* Computers and printers; Internet access for research if necessary.

Development

Prior to starting this project, we suggest that you make a sample decoration. Seeing an example of what they are to do will make it easier for your students to complete this project.

A few days before the project, tell your students that they will make decorations for the classroom and show the one you have created. You might select a theme in advance—Math in Architecture or Pictures of Geometrical Shapes are two examples—or you may leave the theme open to your students. Instruct them to start collecting pictures, 3 by 3 inches in size, on their topic. You may have students bring to class materials from which they may choose pictures, have them find pictures at home, or suggest they check Internet sources. In either case you should have extra sources of pictures on hand. Remind students the day before the project to bring their materials or pictures to class.

- Begin the project by explaining to your students that they will make decorations for the classroom. Mention that the decorations will be three-dimensional octahedra. Once again, show them your example.

- Distribute copies of Student Guide 39.1, and review the information with your students.

- Hand out copies of Data Sheet 39.2, "Constructing a Three-Dimensional Octahedron," and discuss the information. Note that the data sheet provides instructions for making the decorations. Consider having students practice following the directions on plain pieces of paper before they try to make their decorations. This will help them to avoid ruining pictures.

- You may wish to introduce the concept of an equilateral triangle and segment of a circle, as well as review the meaning of diameter of a circle and arcs.

Making three-dimensional octahedra and classroom decorations

- If the construction is too difficult for your class, you may adapt the steps by making several circles whose diameter is 3 inches for students to use as a pattern. You may also cut out and distribute an equilateral triangle you have inscribed to serve as a pattern. In this case, make your patterns out of poster paper or light cardboard, and ask students to trace the circle, place the triangle on it, and trace the triangle. If your computer has software that will allow you to inscribe an equilateral triangle in a circle, showing your students how this is done can be an excellent example of using technology in problem solving.

- Emphasize that students should have eight equilateral triangles, each inscribed in a circle. They will then cut out each circle and fold the segments toward them so that the triangle is flat.

- Point out that the resulting figure will have eight faces, each of which is an equilateral triangle. It is one of the Platonic solids called a *regular octahedron*. You may also wish to discuss Euler's formula: $V - E + F = 2$, where V stands for the number of vertices (in this case, six), E stands for the number of edges (twelve), and F stands for the number of faces (eight).

Wrap-Up

Display the decorations by hanging them in your classroom. Because a large number of decorations will be cumbersome to carry, we suggest you evaluate them as students complete and submit them to you.

Extension

Discuss the other Platonic solids. Students may use the same procedure to construct a regular tetrahedron by using four faces.

STUDENT GUIDE 39.1

Making Three-Dimensional Octahedra and Classroom Decorations

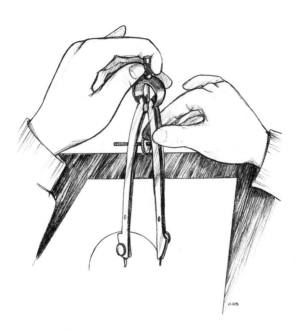

Situation/Problem

You will make a decoration for your classroom by selecting pictures, drawing circles, and inscribing equilateral triangles in them. Your finished decoration will be a three-dimensional octahedron.

Possible Strategies

1. Choose a theme or idea that interests you, and select appropriate pictures.
2. Although you will be working individually, you might work cooperatively with other students in searching for pictures. For example, you may wish to share old magazines. Tell your friends what types of pictures you are looking for, and they may find some; if you find pictures your friends might be able to use, show them.

Special Considerations

- You will need eight pictures for your decoration. Finding more than eight will give you a choice of several. Good sources for pictures are old magazines, catalogues, brochures, greeting cards, wrapping paper, and newspapers.
- If your computer has software with art or drawing capabilities, consider creating pictures for decorations. You may also use clip art if available.

Making Three-Dimensional Octahedra and Classroom Decorations *(Cont'd.)*

- Make sure that your pictures are at least 3 inches by 3 inches.
- Note that the main part of your pictures will be the center of the circles. (If your teacher shows you a model, study it carefully.)
- Review Data Sheet 39.2 carefully. It provides instructions on how to inscribe an equilateral triangle in a circle and construct your three-dimensional octahedron.
- Before beginning your decoration, try making a rough copy using plain pieces of paper. This will give you valuable practice.

To Be Submitted

Your decoration.

Notes

_____

_____

_____

_____

_____

_____

_____

_____

_____

_____

_____

Name _____

Constructing a Three-Dimensional Octahedron

To create your decoration, follow these steps:

1. Inscribe an equilateral triangle in your circle by doing the following:

    Set your compass to make a circle with a 3-inch diameter. Do *not* change the setting as you draw the circle around your pictures.

    The main part of your picture should be at the center of your circle.

    Lightly draw the diameter and label the end points A and B.

    Place the compass point on A and construct an arc, labeling the end points C and D.

    Place the compass point on B and construct an arc, labeling the end points E and F.

    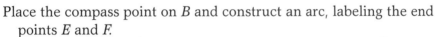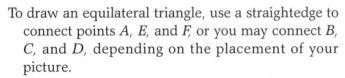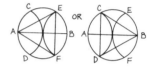

    To draw an equilateral triangle, use a straightedge to connect points A, E, and F, or you may connect B, C, and D, depending on the placement of your picture.

    Make all lines light so that they will not interfere with the picture.

2. Repeat this process for each of the eight pictures you have selected.

3. Cut out the circles.

4. Use the sides of the triangles as lines for folding, and make all folds on the triangle toward you. The folds will result in making flaps, which you will glue together.

5. Take four of the circles, and place glue on two flaps of each circle. (One flap of each circle will not be glued yet.)

6. Place the glued flaps together in pairs so that all four circles are connected. Use big paper clips or binder clips to hold the circles in place until the glue dries. You should now have a three-dimensional object with a circle as the base.

7. Follow the same procedure with the remaining four circles. You should now have two parts of your decoration.

8. Glue the two parts together to complete the decoration. Use big paper clips or binder clips to hold the parts in place until the glue dries.

9. Using a hole puncher, make a small hole in your decoration and attach a thin cord so that your decoration can be hung up.

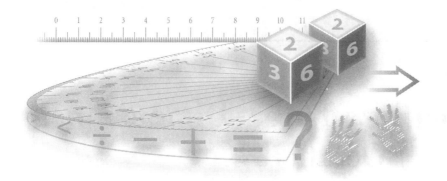

Creating a greeting card for Math Awareness Month

The Joint Policy Board for Mathematics designates April as Math Awareness Month. A unique way to support math education is for your students to create greeting cards that wish others a Happy Math Month.

Goal

Students will work individually to create a greeting card that expresses a thought about mathematics in the context of Math Awareness Month. They will then give (or mail or e-mail) their cards to friends. *Suggested time:* One to two class periods.

Math Skills to Highlight

1. Communicating ideas about mathematics
2. Using technology in problem solving

Special Materials/Equipment

Assorted colors of construction paper; white drawing paper; rulers; compasses; scissors; stencils; felt-tipped pens; markers; colored pencils. If you intend to have students mail their cards, we suggest you use 5- by 8-inch envelopes. *Optional:* Computers and printers; Internet access.

Development

Consider working with your students' art teacher, who may be willing to help create the greeting cards. Before you begin the project, decide if you want students to exchange cards with a friend or mail them to a friend. While having students exchange their cards with a friend in class makes the overall management of the project easier, students enjoy mailing their cards to friends. If the cards are mailed, students may send them to friends in other towns. Mailing will require envelopes and postage.

- Start the project by explaining to your students that they will create a greeting card about math and give (or send) their card to a friend.
- Discuss Math Awareness Month with your students. Note that its purpose is to promote mathematics in schools and communities. By participating in this project, your students will be promoting math by wishing their friends a Happy Math Month.
- Distribute copies of Student Guide 40.1, and review the information with your students.
- Distribute copies of Data Sheet 40.2, "Tips for Creating Math Greeting Cards," and discuss the information on the sheet with your students. Emphasize that the data sheet provides suggestions for making greeting cards.
- If your students have access to computers with software that has the capabilities to create greeting cards, encourage your students to create their greeting cards on computers. Computer-generated cards can be e-mailed rather than sent by postal mail.
- If students are to mail their cards, we suggest that you obtain 5- by 8-inch envelopes. (If your school does not have them, you can obtain them at most office supply stores for a small fee.) These should be sufficient for mailing. Using bigger envelopes will increase postage costs. Of course, you can use smaller envelopes, but these will limit the size of the students' cards. Remind students to create cards that will fit inside their envelopes.

Creating a greeting card for Math Awareness Month

Wrap-Up

Students exchange or address and mail or e-mail their cards.

Extension

Suggest that students e-mail messages about math topics and send them to friends regularly. They might establish e-mail pen pals with whom they correspond about math.

STUDENT GUIDE 40.1

Creating a Greeting Card for Math Awareness Month

Situation/Problem

Working individually, you will create a greeting card containing a thought about math, and exchange or send it to a friend, wishing him or her "Happy Math Month."

Possible Strategies

1. Decide who you will send your greeting card to. Try to include a thought or idea about math that is just right for this person.
2. Think about how you might express an idea about math for Math Awareness Month.
3. Write down as many possible ideas as you can. Then go over them and select the best one.

Special Considerations

- Select an idea that can be expressed on a greeting card.
- Refer to Data Sheet 40.2 for suggestions on how to generate ideas and create a greeting card.
- Consider the design of your card. A simple card may lay flat, or you may design one that folds in half.
- Consider the size of your card. If you intend to mail it to a friend, plan on fitting it inside an envelope. To fit your card inside an envelope, make your card $\frac{1}{4}$ inch less than the envelope's length and width. If you create a card that is to be folded over, its folded dimensions should be $\frac{1}{4}$ inch less than your envelope's dimensions.
- Sketch potential designs on scrap paper first.

Creating a Greeting Card for
Math Awareness Month *(Cont'd.)*

- When you start your actual card, draw your letters and illustrations lightly with pencil. You can go over the lines later with pens or markers.

- If your computer has software with art capabilities, you may want to design your card on your computer.

To Be Submitted

Your finished greeting card.

Possible Ideas

_____

_____

_____

_____

_____

_____

_____

_____

_____

_____

_____

_____

DATA SHEET 40.2

Tips for Creating Math Greeting Cards

There are various ways to create a greeting card, depending on your ideas, interests, and skills. While the following suggestions will help you to make a math greeting card, they cover only the basics. Use your imagination and experiences to create a truly unique card, designed especially for your friend.

Find an Idea

- Think about topics in math you are studying.
- Think about associated ideas or topics.
- Consider your feelings about math.
- Write down any ideas you may have; then pick the one you like best.

Decide on Design

- What color paper will you use?
- Will you express your idea as a short poem? Will you simply write your thought in clear prose?
- Will you use stencils for lettering? Do you know calligraphy? Will you print the words on your greeting card? Will you use script?
- What colors will you use?
- Will you use a computer to design your card?
- Will you draw an illustration? Use computer graphics? Will you use clip art?

The Importance of Unity

- Your letters, illustrations, and overall design should reflect your message. Outstanding greeting cards are those cards whose parts complement each other.

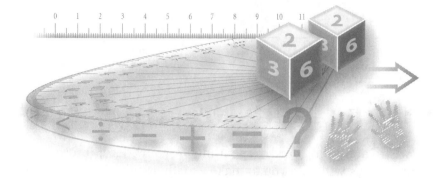

The geometry and art of architecture

All too often students study the concepts of geometry in class but do not connect them to the real world. In this project, students have the opportunity to relate their study of geometry to architecture around the world.

Goal

Working with a partner or in groups of three, students will relate their study of geometry to architecture. They will select a building or structure and identify its geometrical shapes and properties. Each group will then make a poster of that building or structure and label the examples of geometry they found. They will write a brief description of the structure. *Suggested time:* Two periods, although students might need time outside class to complete the project.

Math Skills to Highlight

1. Identifying types of angles
2. Recognizing two- and three-dimensional geometrical shapes
3. Recognizing parallel and perpendicular lines
4. Identifying types of symmetry
5. Using technology in problem solving

Special Materials/Equipment

Reference books on buildings and architecture; poster paper; rulers; felt-tipped pens; markers; colored pencils; stapler and staples. *Optional:* Computers with Internet access for research.

Development

Discuss examples of geometry found throughout your school and classroom. Point out some examples: circular clocks, square tiles, rectangular door frames, corners that are right angles, floor and ceiling that are parallel planes, and rectangular walls that form line segments where they meet.

- Begin the project by explaining that students will work with a partner or in a small group and study the architecture of a building or structure such as the Empire State Building, pyramids, or London Bridge. They will identify as many examples of geometry as they can, create a poster of their building or structure, and write a brief description.

- Consider the level of your class. For students who do not have a strong background in geometry, you may suggest that they look for basic shapes: circles, rectangles, squares, triangles, and so on. For advanced students, suggest they identify the basics as well as arcs, the Golden Rectangle, and specific types of triangles.

- Hand out copies of Student Guide 41.1, and review the information with your students. Note that the student guide contains a list of examples of geometry students should look for. You may wish to add to it.

- Distribute copies of Data Sheet 41.2, "Famous Architecture Around the World." Encourage students to select a building or structure that interests them. This list is certainly not exclusive, and you may wish to open up the project and allow students to select other buildings or structures.

- Provide at least one class period in the library for students to conduct research. Prior to starting this project, consult with your school's librarian and ask her to reserve books on buildings and architecture for your class.

Additional sources where students may find pictures of the buildings they wish to research include encyclopedias, atlases, travel brochures, and magazines (especially *National Geographic* and travel magazines). If possible, have these additional sources available for your students to use. Also suggest that students consult online sources.

- When students do their poster, remind them to draw their building or structure as accurately as possible. They should label as many examples of geometry or geometrical principles as they can.

- Suggest that students write their reports on a computer, which will make the task of revision easier. The reports should be brief, highlighting the geometry represented in their building or structure and providing some background information. The reports should be attached to the bottom of the posters.

Wrap-Up

Display the posters. You might have each group briefly discuss the geometry shown on their posters.

Extension

Have students select a building in town—it may be their own home or even the school—and identify the geometry the building displays.

STUDENT GUIDE 41.1

The Geometry and Art of Architecture

Situation/Problem

You and your partner or group will make a poster of a building or other structure and iden-tify its geometrical shapes and properties. You will also write a brief description of your building or structure.

Possible Strategies

1. List any famous buildings or structures that you and your partner or group members know. Decide if you would like to explore any of these for examples of geometry.

2. Consider choosing a building or structure that is a part of a topic that interests you. For example, if you like medieval history, you might select a castle to study. If you like ancient history, the pyramids may interest you.

3. Review Data Sheet 41.2, and select one of the buildings or structures that is listed.

4. Once you have chosen a building or structure, think about dividing tasks. While all members of the group may work together to identify examples of geometry, you may draw the poster, another student may write the description, and another may edit the writing. All of you may color and label the poster.

The Geometry and Art
of Architecture *(Cont'd.)*

Special Considerations

- You will need to conduct research to find a picture or photograph of the building or structure you have chosen.

- Check books on buildings and architecture, encyclopedias, atlases, history books, geography books, magazines, and similar sources. Magazines such as *National Geographic* often contain photographs of buildings and structures around the world. You might also consult online sources.

- In examining your building or structure, try to find as many examples of geometry as possible. Look for the following:

    Types of angles: acute, obtuse, and right

    Regular polygons such as equilateral triangles and squares

    Other polygons such as right triangles, rectangles, and rhombi

    Circles and semicircles

    Three-dimensional shapes such as prisms, pyramids, cones, domes, and spheres

    Parallel and perpendicular lines

    Symmetry, including reflections, rotations, translations, and combinations

- Draw the building or structure on your poster paper as accurately as you can. Title your drawing, and neatly label the examples of geometry on the poster.

- Write your report using a computer if possible, because this will make the task of revision easier. When you write your description, be sure to include the history or background of your building or structure, as well as a summary of the geometry it represents. Write your description on only one side of the paper (use additional sheets if necessary), and staple it to the bottom of your poster. Be sure to answer the following questions in your description:

    Who designed the building or structure?

    What are its dimensions?

    Where is it located?

    When was it constructed?

    Why was it constructed?

    Is it used today? If yes, how?

To Be Submitted

Your finished poster and description.

DATA SHEET 41.2

Famous Architecture Around the World

The following buildings and structures offer excellent examples of geometry:

Alamo (San Antonio, Texas)

Arc de Triomphe (Paris, France)

Blue Mosque (Istanbul, Turkey)

Castle of El Morro (San Juan, Puerto Rico)

Chrysler Building (New York City)

CN Tower (Toronto, Canada)

Colosseum (Rome, Italy)

Eiffel Tower (Paris, France)

Empire State Building (New York City)

Flatiron Building (New York City)

Geosphere (Lake Buena Vista, Florida)

Great Wall of China (China)

Hagia Sophia (Istanbul, Turkey)

Hancock Tower (Boston, Massachusetts)

Houses of Parliament (London, England)

Huaca del Sol (Moche, Peru)

Independence Hall (Philadelphia, Pennsylvania)

Jefferson Memorial (Washington, D.C.)

Kyongbok Hall (Seoul, South Korea)

Leaning Tower of Pisa (Pisa, Italy)

Lincoln Memorial (Washington, D.C.)

Parthenon (Athens, Greece)

Pei Pyramid at the Louvre (Paris, France)

Famous Architecture Around the World *(Cont'd.)*

Pentagon (Washington, D.C.)

Pont du Gard (Nîmes, France)

Pyramid of Cholula (near Puebla, Mexico)

Pyramid of Khufu (near Giza, Egypt)

Pyramid of the Sun (near Mexico City, Mexico)

Schönbrunn Palace (Vienna, Austria)

Sears Tower (Chicago, Illinois)

Shwe Dagon (Rangoon, Myanmar)

Space Needle (Seattle, Washington)

St. Basil's Cathedral (Moscow, Russia)

St. Peter's Square (Vatican City, Rome)

Sydney Opera House (Sydney, Australia)

Taj Mahal (Agra, India)

Temple of Warriors (Yucatán, Mexico)

Tower Bridge (London, England)

Transamerica Building (San Francisco, California)

United Nations Building (New York City)

Washington Monument (Washington, D.C.)

U.S. Capitol Building (Washington, D.C.)

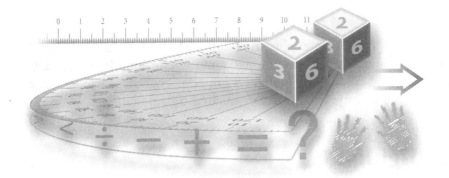

Designing a quilt pattern

People have used quilts for at least a few thousand years. The ancient Russians, Chinese, and native civilizations of Central America wore quilted clothing for warmth. When the Crusaders returned to Europe, they brought home the idea of quilted fabrics, which they learned from the Saracens, who wore quilted shirts. The Europeans soon adapted quilts for undergarments and sleepwear. When the Dutch and English colonists settled the New World, they were kept warm by the quilts they laid across their beds. Today quilt making is as much art as craft. The elaborate patterns quilts exhibit inspire both admiration and fascination. Although your students will not actually make quilts for this project, they will have the opportunity to design original quilt patterns.

Goal

Working individually, students will create and color a one-patch quilt design based on the regular hexagon. A one-patch quilt is made by using only one geometrical shape that is repeated throughout the quilt. (The hexagon is a good choice, because it can be cut in a variety of ways, such as isosceles trapezoids, rhombi, isosceles triangles, equilateral triangles, and kites.) Students will choose a design, draw the pattern on Worksheet 42.3, and color it to highlight a quilt design. *Suggested time:* One to two class periods, with some time possibly needed outside class.

Math Skills to Highlight

1. Recognizing the properties of regular polygons, particularly the regular hexagon
2. Recognizing the properties of an equilateral triangle
3. Measuring angles
4. Finding the sum of the measures of interior angles of triangles and quadrilaterals
5. Identifying lines and points of symmetry
6. Identifying congruent figures
7. Using technology in problem solving

Special Materials/Equipment

Rulers; protractors; felt-tipped pens; colored pencils; markers; pictures of quilts or a real quilt. *Optional:* Reference books about quilts and quilting; computers and printers; Internet access for research.

Development

If you have a quilt, bringing it in to class to show your students is an excellent way to generate interest in this project. If you cannot bring in a quilt, try to obtain books about quilts from the library. You might also suggest that students visit Web sites devoted to quilts and quilting, where they will find many examples.

- Begin the project by explaining that each student will create a one-patch quilt design, based on a regular hexagon. After completing their designs, students will color them to enhance their quilt patterns.
- When you show pictures of quilts, ask students to identify some of the figures they recognize on your sample quilts, and point out how they are pieced together to achieve an overall effect.
- Distribute copies of Student Guide 42.1. Review the information with your students, and emphasize the need for careful measuring and use of congruent figures.

- Hand out Data Sheet 42.2, "Creating a One-Patch Quilt Design," and discuss the suggestions and shapes.

- Remind students that they may use only one shape since they are designing a one-patch quilt.

- Hand out at least two copies of Worksheet 42.3, "Quilt Design," to each student. Have extra copies ready for students who wish to try various patterns or who make mistakes with their designs. Mention that the worksheet has a border, which is not part of the design, and suggest that students use the dots for guidelines. The designs will stretch beyond the border, but students should continue to draw as much as they can in order to fill the worksheet.

- If students have access to computers with software that has art and drawing capabilities, suggest they try designing their patterns on computers.

- Explain that the use of color can enhance their designs. Note that the selection of colors may enhance or detract from their designs. Particular colors can make some shapes stand out while others may make some shapes difficult to see.

- Once the designs are finished, ask students to measure the angles and test a conjecture about the sum of the measures of the interior angles of a polygon. You may wish to introduce the formula $(n - 2)180$, where n stands for the number of sides.

- Depending on the abilities of your class, you may wish to discuss the properties of regular polygons and congruent figures. Another aspect of this project is the important role of symmetry. Symmetry with respect to a point or a line may also be introduced or reinforced.

Wrap-Up

Display the quilt designs on the bulletin board.

Extension

Invite a quilter to the class with examples of her work. Have her explain quilting to your students.

Designing a Quilt Pattern

Situation/Problem

You will create a design for a one-patch quilt based on a regular hexagon. A one-patch quilt is made by using only one geometrical shape that is repeated throughout the quilt. You will not have to select fabrics or sew; you only have to make the design on the worksheet. You should color your design to highlight its pattern.

Possible Strategies

1. View examples of quilts and study their patterns. Consult both print and online sources.

2. Review Data Sheet 42.2 to help you create a possible design.

3. Use Worksheet 42.3 to draw a rough sketch of your design and add color. This will help you to better visualize what the finished design will be like. Revise your initial design as necessary.

Special Considerations

- Draw in pencil. Use a straightedge.

- You are making a one-patch quilt so you may use only one shape. Be sure to consult Data Sheet 42.2 for details.

Designing a Quilt Pattern *(Cont'd.)*

- Use Worksheet 42.3 to help you make your design. Or consider designing your pattern on a computer. You will need software that allows you to draw and work with figures and shapes.

    The border around the page will not be part of your design.

    Use the dots on the worksheet to help you draw your design.

    Start near the middle of the grid, and draw your design, repeating until the grid is nearly filled. Your design will stretch to the border.

- Once your design is finished, color it. Keep in mind the following suggestions:

    Choose colors you like.

    Decide what you wish to be highlighted on your design. Choose colors that will help this part of your design to stand out.

    Select colors that are pleasing to the eye and that compliment each other. For example, purple and a close shade of blue may run together and weaken any contrast. Green and yellow will provide a contrast.

    Color the border.

To Be Submitted

Your quilt design.

DATA SHEET 42.2

Creating a One-Patch Quilt Design

To create a quilt design based on a regular hexagon, you may use one of the following patterns or create one of your own. Remember that once you select a shape, it is the only one you can use in your design.

A regular hexagon can be divided to form:

- Two isosceles trapezoids

- Three 60°–120°–60°–120° rhombi

- Six isosceles triangles

- Six equilateral triangles

- Six kites

Once you have selected the shape you would like to use, repeat this shape to form other geometrical figures. If you can find another way to divide a regular hexagon, you can use this figure to design your quilt.

There are several possible quilt arrangements.

Quilt Design

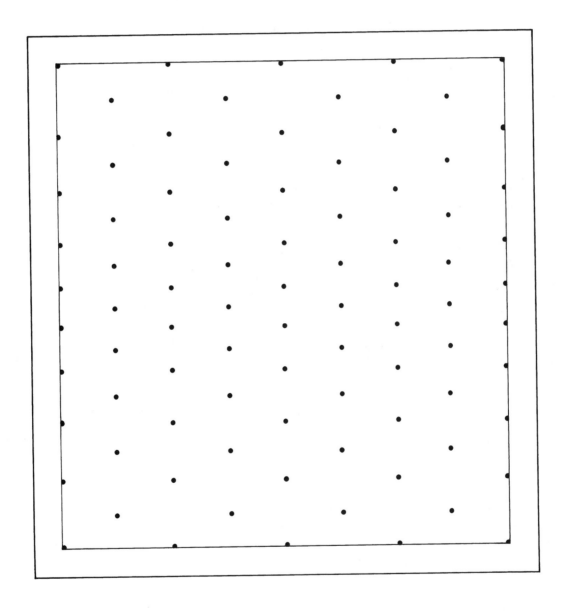

Math and Sports and Recreation

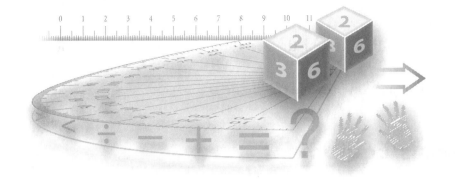

Choosing a membership plan at a health club

Students and adults are constantly forced to make choices based on a variety of factors. This project, which requires students to evaluate the membership plans of a health club, provides a glimpse of one of those real-life choices.

Goal

Working individually, students will pretend that they are about to join a health club. They are to analyze various membership plans and select the one that is right for them. They will write an explanation detailing their reasons for choosing the plan they did. *Suggested time:* One to two class periods.

Math Skills to Highlight

1. Using estimates to solve problems
2. Determining the best options

3. Comparing and contrasting the costs and benefits of different membership plans

4. Using technology in problem solving

Special Materials/Equipment

Calculators for comparing costs. *Optional:* Computers and printers.

Development

Discuss problem solving as a basis for making intelligent decisions. This is particularly true in many common activities such as choosing a car, buying admission plans to amusement parks, and vacation planning. Although cost is usually a factor, it is not the only factor. What seems to be a good deal in terms of price may not in fact be very good if you are paying for a lot of extras you do not need. Conversely, a basic plan that does not offer any extras may present the best value out of several plans because it contains the things people need the most.

- Begin this project by asking students if any of them belong to a health club. If you teach high school, particularly the upper grades, it is likely that some of your students do. You might also ask how many of their parents belong to health clubs. Briefly discuss that a health club is a place that offers activities for keeping fit. Such activities might include weight-training and exercise equipment, aerobics classes, tennis, volleyball, basketball, and even swimming.

- Distribute copies of Student Guide 43.1, and review the information with your students. Emphasize that students are to select a membership plan that satisfies their needs at the lowest cost.

- Note that in comparing plans, students may have to estimate the number of times they will be able to use the club.

- Hand out copies of Data Sheet 43.2, "Power Health Club Membership Plans." Review the sheet with your students. Be sure to point out the different plans, and note that each has different features. They should be especially careful not to buy a membership that contains a lot of options that they will not need. This increases the cost with very little benefit to them.

- Remind students that they are to write an explanation of why they chose the plan they did.

Wrap-Up

Ask volunteers to explain what plan they chose. Collect the written explanations and display them.

Extension

Ask students to speculate on why the various membership plans are set up the way they are. Why might a health club offer yearly commitments or a pay-as-you-go plan? Why might it offer several different plans? What advantages might a health club gain by offering various plans to people? What are the advantages, if any, to the potential members?

STUDENT GUIDE 43.1

Choosing a Membership Plan at a Health Club

Situation/Problem

Imagine that you have decided to join a health club. You visited the club and asked for information about membership plans. When you review the information, you find that there are several plans and want to choose the plan that best suits your needs, but at the most reasonable cost. After making your selection (from Data Sheet 43.2), write an explanation, offering your reasons for choosing the plan you did.

Possible Strategies

1. Think about your needs. What types of activities would you like to do at a health club: weight training, exercising using different kinds of equipment, aerobics, basketball, tennis, or volleyball, for example?
2. Examine the various plans to see which ones offer what you want most.
3. Consider the times you could go to work out. Are all of the activities you want to participate in available when you can go?
4. Compare the plans, and evaluate the costs of each.

Special Considerations

- Review Data Sheet 43.2, which contains several membership plans from which you can choose.
- Carefully consider whether you would be satisfied participating in only your favorite activities. Do you want the privilege of taking part in everything the club offers?

Choosing a Membership Plan
at a Health Club *(Cont'd.)*

- Check the off-hour rates. If the activities you are most interested in are offered then, you may be able to save on costs.

- To compare the costs per visit of different yearly plans, estimate the number of times per month you expect to work out and multiply by 12. This will give you the total number of visits for the year. Now divide the cost of the plan by the number of visits. Your answer will be the cost per visit. Compare that to the cost of "pay as you go."

- Some activities require partners (tennis), and others require teams (volleyball and basketball). Does the club offer leagues, which will provide you with partners or teams? Or will you have to join with friends? If you were to join with friends, would they be able to work out when you could?

- After selecting your membership plan, write an explanation detailing your reasons for choosing this plan over the others. Be prepared to orally explain your reasons.

- Write your explanation on a computer, which will make the task of revision easier.

To Be Submitted

Your explanation.

Notes

Name _____

Power Health Club Membership Plans

Gold Membership: One-year membership, $919.00. Unlimited use of facilities during times the club is open. Activities include weight training, use of exercise equipment, aerobics classes, tennis, basketball, and volleyball.

Two-for-One Gold: Join with a friend. One-year membership for two: $1,399.00. Includes same privileges as the Gold Membership.

Silver Membership: One-year membership, $699.00. Unlimited use of weights, exercise equipment, and aerobics classes during times the club is open.

Pay-as-You-Go Membership: $18.00 per visit for use of the weights, exercise equipment, and participation in aerobics classes during times the club is open. $25.00 per visit for tennis, basketball, or volleyball during non–prime time and $30.00 during prime time. (Prime time is between 7:00 P.M. and 9:00 P.M., Monday through Friday.) Note: Leagues are available for tennis, basketball, and volleyball. A one-time registration fee of $15.00 is required for joining each league.

Aerobics Membership: One-year membership, $399.00. Includes unlimited participation in aerobics classes.

Off-Hours Membership Plan: One-year membership, $359.00. Unlimited use of facilities between 2:00 P.M. and 6:00 P.M. every day. (Activities are the same as the Gold Membership.)

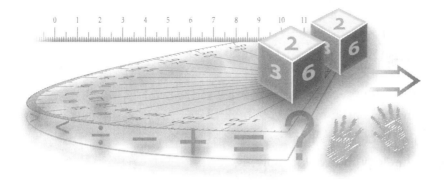

Equipment for the school's workout room

Ask students what their favorite subject in school is, and many will answer gym. Of course, most of these students are offering this answer only as a joke, but gym nevertheless remains a popular class. In this project, students have the opportunity to imagine that they are to select equipment for a new workout room that has been added to their school gym.

Goal

Working in groups of three or four, students are to assume that they have been asked to help select equipment for a new workout room. They are given a $6,000 budget that they cannot exceed. They are to make their selections and complete an order form for equipment, as well as share their choices with the class. *Suggested time:* Two to three class periods.

Math Skills to Highlight

1. Using estimation
2. Making choices based on budget constraints
3. Maintaining accurate tallies
4. Using technology in problem solving

Special Materials/Equipment

Calculators. *Optional:* Catalogues containing workout equipment; computers with Internet access for research.

Development

To focus this project, think of the workout room containing equipment such as rowing machines, cross-country skiing machines, treadmills, and exercise bikes. If you are not familiar with such equipment, consult with your students' physical education teacher. Prior to assigning this project, obtain old catalogues and advertisements containing workout equipment that students can use to find the machines they might like to buy. Data Sheet 44.2, "The Cost of Exercising," also contains equipment and prices.

- Begin this project by telling your students to imagine that a workout room has been added to their school's gym, and they can use it during free periods and after school. Groups of students have been given the opportunity to choose the equipment.

- Distribute copies of Student Guide 44.1, and review the information with your students. Note that each group may spend up to $6,000. They should try to buy equipment that will be popular among the greatest number of students.

- Distribute copies of Data Sheet 44.2, "The Cost of Exercising." Review the material with your students, and explain that this data sheet concentrates on five major areas of exercise: cycling, weight training, rowing, walking or running, and cross-country skiing. Students may choose equipment from the data sheet.

- If you have obtained catalogues or advertisements containing workout equipment, allow students to select equipment from these sources as well as from the data sheet. You may also suggest that students visit Web sites that provide information on exercise equipment. Students may include equipment and machines other than the kinds listed on the data sheet; for example, jump ropes and stair climbers.

- Some students may have workout equipment at home. Encourage them to share their knowledge and experience of these machines with the other members of their group.

- Emphasize that the many types of equipment should be carefully considered. Suggest that students discuss the value of the various products, and choose what they feel will help to create a workout room that most students would like to use.

- Distribute copies of Worksheet 44.3, "Gym Equipment Order Form." Review the worksheet with your students, and explain what the different categories mean. After selecting the equipment they would like to purchase, each group is to fill out the form, being sure to include all the necessary information.

- Remind students to select a spokesperson to share their selections with the class. Each group should be able to justify its choices.

Wrap-Up

A spokesperson for each group presents its selections of equipment. Students should also submit their worksheets.

Extension

If students had an additional $3,000 to spend, what else would they purchase? Would they buy more of the equipment they already chose or different items?

STUDENT GUIDE 44.1

Equipment for the School's Workout Room

Situation/Problem

Imagine that your group has been asked to select equipment for a new workout room that has been added to your school's gym. This workout room, which will be available to students during free periods and after school, will contain equipment such as rowing machines, cross-country skiing machines, treadmills, and exercise bikes. The school board will provide $6,000 for you to buy the equipment. You cannot exceed that amount, and any money you do not spend will be returned to the board.

Possible Strategies

1. Brainstorm to list some types of equipment you might consider purchasing. The equipment you consider should appeal to as many students as possible.

2. Divide the tasks for the project. While all group members should be involved with choosing the equipment, one may be responsible for writing down the reasons for choosing specific equipment or models over others, another may keep track of estimated costs, a third may be responsible for completing Worksheet 44.3, and the fourth may assume responsibility for presenting the group's selections to the class.

3. As you consider various types of equipment, keep a running estimate of your total costs. If two or three pieces of equipment consume most of your $6,000 allotment, you will not be able to buy much more.

Equipment for the School's
Workout Room *(Cont'd.)*

Special Considerations

- Use Data Sheet 44.2 to review and select equipment.
- If you have catalogues of workout equipment or advertisements containing such equipment, use them along with the data sheet to make your selections. You might also check the prices of equipment on the Internet. Conducting a search using the terms "fitness equipment" or "exercise equipment" will likely result in several helpful sites.
- Carefully weigh the pros and cons of each piece of equipment you are considering. Choose equipment that you believe will benefit the most people.
- After you have decided on which equipment your group would like to purchase, complete Worksheet 44.3. Be sure that you total the prices accurately and include reasons (justification) for selecting the equipment you did.
- Do not calculate sales tax. Since schools are nonprofit institutions, most purchases they make are not subject to sales tax.
- There is no shipping or handling fee. (Many suppliers waive shipping and handling for large orders.)
- Appoint a spokesperson to present your selections to the class. Be able to explain why you have made these choices.

To Be Submitted

Your worksheet.

Notes

_____

_____

_____

_____

_____

_____

_____

_____

The Cost of Exercising

Listed below are several popular types of exercise equipment, their benefits, and prices of specific models:

Cycling

Equipment: stationary bike

Benefits: excellent aerobic exercise; tones and strengthens legs and circulation

- Model: *Basic Rider.* Standard unit, $349, includes built-in timer, odometer/speedometer.
- Model: *Premium Rider.* $549, includes controls to monitor distance, speed, and resistance.
- Model: *Rider 2000.* Top-of-the-line unit, $1,295. Includes built-in timer, odometer/speedometer; automatically adjusts the resistance on leg muscles during cycling.

Weight Training

Equipment: Free weights or multistation weight unit

Benefits: Strengthens, tones, and shapes every muscle in the body

- Model: *Barbell Starter Set.* $229. Basic barbell set includes solid steel bar and complete set of weights.
- Model: *Standard Bench Press.* $279. Basic weight-training bench. Includes weights.
- Model: *Inclined Exerciser.* $669. Features adjustable exercise bench; allows thirteen basic exercises. Weights included.
- Model: *Power Weight I.* $1,879. Features six exercise stations for both upper- and lower-body workouts. 170-pound weight stack.
- Model: *Power Weight II.* $2,995. Features six exercise stations and more than 100 different exercises. Includes 260-pound weight stack.

Rowing

Equipment: Rowing machine

Benefits: Strengthens cardiovascular system, tones overall body

- Model: *Easy Row.* $399. Uses hydraulic pistons to give the feel of rowing; ten resistance settings.
- Model: *Advanced Row.* $879. Simulates actual rowing, fully electronic; twenty resistance settings; has pull strap instead of oars.

The Cost of Exercising *(Cont'd.)*

Walking and Running

Equipment: Motorized or nonmotorized treadmill

Benefits: Excellent for strengthening the cardiovascular system; tones entire lower body.

- Model: *Walk-Well.* $839. Nonmotorized; has digital timer and speedometer. Elevation may be adjusted.
- Model: *Powertread.* $2,549. Motorized; computerized control panel shows speed, running time, distance, and pace per mile.

Cross-Country Skiing

Equipment: Cross-country ski machine

Benefits: Strengthens and tones all muscle groups

- Model: *Pro Ski.* $579. Pulley system simulates movement of cross-country skiing. Adjustable tension knob.
- Model: *Pro Ski Plus.* $769. Includes same features as the Pro Ski but also includes electronic timer and speedometer.

Name _____

Gym Equipment Order Form

Item Number	Model Name	Justification for Ordering	Price Each	Total Price

Total = _____

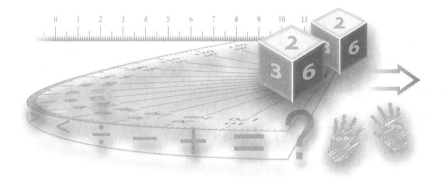

Comparing sports superstars

Most students have favorite sports and are familiar with the stars of that sport. While many athletes are awarded superstar status by the media, their actual statistics and value to their teams may not be as good as their reputations. As students compare superstars in their favorite sport for this project, it is likely they will be surprised by what they find.

Goal

Working in pairs or groups of three, students will select a sport and at least five athletes generally considered to be superstars in that sport. The students will research the careers of these athletes, compare their statistics, and then determine who is the best. They will create graphs, charts, or tables to represent their findings, and orally share their results with the class. *Suggested time:* Two class periods.

Math Skills to Highlight

1. Researching, analyzing, and comparing statistics
2. Understanding *average,* a term often used in sports statistics
3. Making a determination based on statistics
4. Creating graphs, charts, or tables
5. Using technology in problem solving

Special Materials/Equipment

Poster paper; rulers; felt-tipped pens; markers; reference books on sports stars and player statistics. *Optional:* Computers and printers; Internet access for research; an overhead projector and transparencies, PowerPoint, and interactive whiteboard for presentations.

Development

Prior to beginning this project, ask your school librarian to reserve books on sports. Web sites may also provide excellent information. Still another source of information is sports cards: baseball, football, basketball, and hockey cards offer plenty of information about players. You might suggest that students who have card collections review them for players and statistics.

- Begin this project by explaining that every sport has its so-called superstars. Many of these athletes enjoy their status because of flamboyant styles, the ability to make great plays under pressure, or simply media hype. Some superstars, however, do not have great statistics. Other players in their leagues may have better stats, but because they contribute to their teams' success without much fanfare, their real value is overlooked.

- Explain that each pair or group of students is to pick a favorite sport, and choose at least five athletes in it who are considered to be superstars. (Obviously this is subjective, but most fans know who the top players in their favorite sports are.) Students are to research the statistics that prove the achievement of these athletes and decide who is the best.

- Encourage students to choose more than five athletes if they wish. You might also suggest that they compare the statistics of the superstars to those of other players.

- Distribute copies of Student Guide 45.1, and review the information with your students. Note that they may select any sport for this project, but the more popular ones, such as baseball, basketball, football, hockey, tennis, and soccer, will be easier to research.

- Schedule a period in the library for students to conduct research, and encourage them to do additional research on their own using both print and online sources.
- Suggest that students develop criteria by which to compare athletes. Point out that the criteria should be consistent and that athletes should be compared according to the same categories.
- Distribute several copies of Worksheet 45.2, "Sports Superstars," to each group. Recording players' statistics on the worksheet will make comparison easy.
- Review the word *average* with your students. Many sports statistics are expressed as averages. For example, in basketball, a player's scoring average is the total number of points he or she scores in a season, divided by the number of games he or she plays. Averages are an excellent means of comparison. A basketball player who averages 25.4 points per game is a much better scorer than one who averages 8.5. For some sports, you might feel that it is necessary to review decimal values with your students because some averages are expressed as decimals.
- After students make their determination of which athlete is the best, they should create graphs, tables, or charts to illustrate their findings.
- Remind each group to select a spokesperson to present the group's findings to the class.
- Consider having students use an overhead projector, PowerPoint, or an interactive whiteboard in their presentations.

Wrap-Up

Oral presentations. Display the graphs, charts, and tables.

Extension

Instruct students to compare the salaries of some of the highest-paid players in sports. They should calculate the player's pay per game or event, and discuss if these athletes are paid fairly, or too much, or too little. Set up discussion groups of six to eight students for this discussion.

STUDENT GUIDE 45.1

Comparing Sports Superstars

Situation/Problem

You and your partner or group are to select a sport and at least five athletes in that sport who are generally considered to be superstars. You are to research and compare their career statistics that show their contributions to their teams. You will then determine if these athletes are truly superstars. After making your determination, you are to create graphs, tables, or charts to support your conclusions. You will present your conclusions to the class.

Possible Strategies

1. Decide which sport you and your partner or group would like to research.
2. Discuss what makes a superstar in the sport you have chosen. List specific criteria or examples.
3. List at least ten athletes you believe are superstars in this sport. Narrow this list to five to seven.

Comparing Sports Superstars *(Cont'd.)*

Special Considerations

- Research the statistics of the athletes you have selected. Books on specific sports and athletes, references on sports statistics, almanacs, and books of records will likely provide good information, as will Web sites devoted to sports and athletes. Baseball, football, hockey, and basketball cards are also good sources of facts.

- Be sure to compare the athletes you have selected according to specific criteria. For example, in baseball, you might compare players on:

    Batting average

    Runs batted in (RBIs)

    Home runs

    Triples

    Doubles

    Total number of hits

    Total number of runs scored

    Fielding percentage

    Games played

    You would need to compare starting pitchers according to different criteria, such as won/lost record, ERA (earned run average), strikeouts, and bases on balls.

- Use Worksheet 45.2 to record and help you compare the statistics of players. Using one sheet per player will make it easier to compare players.

- Consider comparing the superstars to some other players. Are there other players in your sport who are not usually thought of as being superstars but who have better stats than some superstars?

- Based on your comparison and analysis of statistics, determine which athletes are the true superstars.

- Create graphs, tables, or charts to support and illustrate your conclusions. For example, you might use graphs to compare the stats of the superstars in specific categories. If possible, create your support materials on computers.

- Select a spokesperson to present your conclusions to the class.

To Be Submitted

Your graphs, charts, or tables.

Sports Superstars

Fill in the superstar's name, and label the categories for statistics across the top. *Example:* For a basketball player you might wish to record games played, points scored, scoring average, rebounds, and steals. Listing the years along the left will provide you with useful statistics for this player, which you can then compare to others.

Name of Superstar: _____

Year					

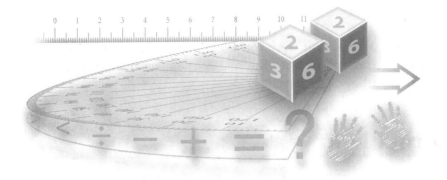

Math and the big game

Many students are enthusiastic sports fans or have at least a moderate interest in sports. Most are familiar with major sport events such as the World Series and Super Bowl. This project is for the sports buffs in your class. Students are required to watch a major sport event and record examples of statistics that are offered throughout the game. Your students will quickly realize that without math, the game would lose much of its appeal. Since this project will interest primarily students who enjoy sports, you might wish to make it a voluntary or extra credit assignment.

Goal

Working individually, students will watch a major sport event such as a game of the World Series, the Super Bowl, or the National Collegiate Athletic Association (NCAA) Men's Basketball Finals or the Women's Basketball Finals. They are to record the statistics offered throughout the game and note how the use of math helps to enrich understanding of the action. When they are finished, they are to select one or two statistics that they felt were most helpful to viewers and be ready to share their opinions with the rest of the class during a discussion. *Suggested time:* Two partial class periods.

Math Skills to Highlight

1. Using statistics to understand relationships in sports
2. Gaining an appreciation of the importance of mathematics in sports

Special Materials/Equipment

None.

Development

Because big games will generate the most interest among students, consider assigning this project during the World Series (seventh game if possible), a major college football bowl game, the Super Bowl, the championship game of the NCAA basketball finals, or another big game. The big games are also the ones in which sportscasters flash countless statistics on the screen for viewers. (After this project is done, your students will probably be surprised at just how many statistics are offered.) Plan on completing the project in two partial periods: a fifteen- to twenty-minute session to introduce it and another fifteen- to twenty-minute session after the game for the discussion. If you assign the project as an optional activity, you may conduct the discussion with the participants at the back of the room while the rest of the class works on the day's homework.

- Begin this project by telling your students that math has a major role in sports. Aside from the obvious need of numbers to keep score, games are played on fields and courts of specific dimensions, many rules are based on numbers (for example, baseball's three strikes and you're out), and equipment must conform to precise sizes. Also, players are compared by statistics. The quarterback who completes 60% of his passes is better than the one who completes 45%. It is likely that your students have watched hundreds (maybe even thousands) of games without thinking about the significance of mathematics.

- Tell students what game you want them to watch. Explain that they are to record as many statistics displayed on the TV screen as they can, noting their purpose to the action of the game. They should try to answer the questions: Why is each statistic displayed? What does it add to watching the game?

- Distribute copies of Student Guide 46.1, and review the information with your students.

- Emphasize that students are to select one or two statistics that they found most helpful in adding insight or background to the game. They should be able to discuss their opinions at the conclusion of the project.

- Make several copies of Worksheet 46.2, "Tracking the Stats," available to your students. This worksheet is designed to help them record statistics and note their impressions. Each student should have at least three sheets. If they need more space, instruct them to use the back of the sheets or a separate sheet of paper.

Wrap-Up

Conduct a discussion in which students share their opinions about the statistics used during the game.

Extension

Expand the discussion to this question: Might the overuse of statistics detract from a game?

STUDENT GUIDE 46.1

Math and the Big Game

Situation/Problem

You will watch a major sports event on TV, and record as many examples of the use of statistics as you can. As you record the stats, you will try to identify their purpose and decide whether each is effective. After the game, you will select one or two stats that you found most helpful to your understanding or appreciation of the game. On returning to class, you will share your opinions during a discussion.

Possible Strategies

1. Watch the game with several copies of Worksheet 46.2 handy. When a statistic appears, write it down. If you wait, you might forget what it was.

2. Include an example of each statistic you record. This will help you to recall them clearly.

Math and the Big Game *(Cont'd.)*

Special Considerations

- Label your statistics according to quarters, innings, or periods. This will make it easier to identify specific stats during the follow-up discussion.

- Try to record as many statistics as you can, but do not worry if you miss a few. Sometimes stats are flashed on the screen too quickly to copy.

- Try to identify the purpose of each statistic. For example, in football, if one of the teams playing leads the league in defense, what is the significance of that? Obviously it suggests that this team should be able to hold its opponent to a low score. If the opponent has already scored three touchdowns midway through the first half, the league's leading defense is not playing well. What might that mean for the rest of the game?

- Some statistics are more helpful to understanding a game than others. Try to identify ones that are the most helpful to you. Note them on your worksheet. Be ready to share your opinions and supporting reasons with the class.

- If you run out of worksheets, use the back or a separate sheet of paper.

- If you watch the game with a friend who is also doing this project, be sure that each of you formulates his or her own opinions about the statistics provided during the game.

To Be Submitted

Your worksheets.

Tracking the Stats

Use this sheet to record the statistics shown during a major sports event.

Name of Event: _____

Statistic	Purpose and Effectiveness

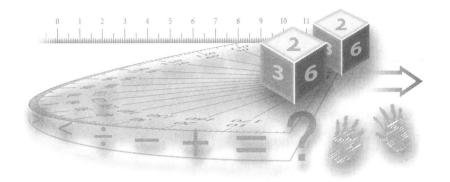

Your unique exercise program

Most adolescents realize the value of exercise. Books and magazine articles that promote fitness, as well as advertisements for exercise equipment, health club memberships, and exercise videos, are common. While many students exercise regularly through sports and dance, many others express an interest in developing a personal exercise program. That is what this project encourages them to do.

Goal

Students will work in pairs or groups of three to design an exercise program. They will explain the program they develop to the class and write a brief description of it. On completion of the project, students will be encouraged to commit themselves to their exercise programs. *Suggested time:* Two to three class periods.

Math Skills to Highlight

1. Using a stopwatch to take a pulse

2. Determining a target heart rate

3. Finding the percentage of a number

4. Rounding to the nearest whole number

5. Using technology in problem solving

Special Materials/Equipment

Stopwatches; calculators; fitness magazines and books. *Optional:* Computers with Internet access.

Development

Ask your students what they do for exercise, and their answers may run from participating in sports, dance, and gymnastics to simply working out at home or in a gym. Ask students why they exercise, and you will likely receive equally diverse answers, from keeping healthy to staying slim to building muscles.

A cautionary note: Before starting this project, check with your school's nurse and your students' gym teacher regarding any students who, because of medical reasons, are restricted from participating in physical activities. Most students who are restricted in some way will tell you. Occasionally there is one who will not because he or she wants to participate. Privately remind students who are restricted not to attempt any workouts that violate their doctor's advice. Because of its focus on fitness, you may want to invite your students' gym teacher to work with you on this project.

- Begin this project by explaining that students will work in pairs or groups of three to develop an exercise program. In setting up partners or groups, consider allowing friends to work together. Friends often have similar interests for exercising and will more likely work out together after they have developed their program.

- Distribute copies of Student Guide 47.1, and review the information with your students. Point out that the guide contains a list of exercises students may consider in developing their programs. There are many more, of course, and you should encourage students to consult reference books and magazine articles about fitness and exercise. Students may also consult online sources.

- Discuss the value of aerobic exercises that cause the heart to beat faster for an extended period (at least twenty minutes). Aerobic exercises strengthen the cardiovascular and circulatory systems and promote general fitness. Jogging, rope skipping, cross-country skiing, swimming,

344

ice-skating, basketball, hiking, dancing, and brisk walking are good examples of aerobic exercises.

- Hand out copies of Data Sheet 47.2, "Your Training Range." Review the information with your students, and discuss the importance of exercising within the target zones as noted on the sheet. If necessary, review how to find the percentage of a number and rounding to the nearest whole number, which students will need to do to determine their own training range.

- Plan to spend a period in the library so that students may conduct research on various exercises. Prior to beginning this project, ask your librarian to reserve books and magazines on fitness and exercise. The Internet is also a good source of information.

- Emphasize the importance of developing an exercise program that is reasonable and suited to the person. *Also note this caution:* Students who do not exercise regularly should consult with their doctor before embarking on any exercise program. During exercise, if students feel that they are becoming fatigued or light-headed or are having trouble catching their breath, they should slow down.

- Encourage students to commit themselves to their exercise routine.

- Note that charting one's progress can be an important part of an exercise program. Although this is not a requirement of the project, encourage your students to keep a record of the progress they make in their exercise program. Recording progress enables students to see how they are improving and provides continuous motivation. An exercise chart may be little more than a dated log that shows the progress toward goals. Mention that it often takes time to achieve fitness goals, and students should not be discouraged by what may seem to be slow progress in the beginning of an exercise program.

- Remind students to write a description of their exercise plan, and be prepared to share their plan orally with others.

Wrap-Up

Conduct a discussion in which students describe their exercise plans. Also, display their written descriptions.

Extension

About three months after the completion of the project, ask your students to evaluate the exercise plan they developed. Have they met their goals? Have they increased the level of activity? Have they expanded their exercise plan to include other forms of exercise? Have they given up? If yes, why?

STUDENT GUIDE 47.1

Your Unique Exercise Program

Situation/Problem

You and your partner or group will develop an exercise program. You will explain your program to the class, and write a brief description of it. You are encouraged to commit yourself to your program.

Possible Strategies

1. Choose exercises that you enjoy.
2. In developing your program, specify the types of exercises, the number of workouts per week, and the length of time for each workout.

Special Considerations

- There are many activities you might include in an exercise program. The following is just a partial list. Consult exercise and fitness books, as well as Web sites, for more:

    Basketball

    Baseball

Your Unique Exercise Program *(Cont'd.)*

Softball

Weight training

Volleyball

Jogging

Dancing (especially aerobic dancing)

Calisthenics (such as sit-ups, push-ups, squats)

Exercise videos

Ice-skating

In-line skating

Cycling

Hiking

Walking (briskly)

Swimming

Hockey (field, street, or ice)

Soccer

Tennis

Cross-country skiing

Downhill skiing

Snowboarding

You may find it necessary to research some of the activities you are considering.

- Most fitness experts suggest a workout plan of at least three sessions a week for at least twenty minutes of steady activity (not including a warm-up and cool-down period). A program of three to four sessions a week of forty-five to sixty minutes per session of steady exercise is considered to be a vigorous program.

- You might develop a program in which you alternate some exercises with the seasons. For example, if you live in a part of the country that receives a lot of snow in the winter, you might substitute cross-country skiing for jogging (which you would do during the late spring, summer, and early fall). Similarly, if you have access to ice in the winter, you might substitute ice-skating for in-line skating. If you live near the shore or belong to a community pool, you might swim in the summer.

Your Unique Exercise Program *(Cont'd.)*

- As you consider which exercises to include in your program, also consider your fitness goals. For example, your fitness goals might include:

    Losing or gaining weight

    Reducing or increasing your measurements

    Improving your strength, energy, or stamina

    Reducing stress

    Improving your general health and sense of well-being

- A necessary part of any exercise program is safety. Choose exercises that are suited to you. Avoid selecting exercises that are too difficult or too demanding. Be willing to start at low levels of exertion and build your endurance. For example, if you seldom jog, do not select a three-mile run as the most important part of your exercise program. Start with a quarter-mile walk or slow jog around the school track and gradually improve your stamina.

- If you have not worked out regularly, are recovering from an illness, or suffer from a serious condition, consult your doctor before beginning any exercise program. In addition, keep the following points in mind:

    Always warm up before exercising by stretching. Going lightly through the motions you will use while exercising is a good way to warm up.

    Always cool down after exercise by walking around for a few minutes. This allows your body to ease back to normal. Never just plop down on a chair while your body is still breathing hard.

    Drink plenty of liquids, preferably water.

    If you feel you are becoming fatigued or light-headed, or experience pain, stop exercising. These are signs you are pushing your body beyond its safety zone.

- A good guide for safe exercise is to work out within your training range. This range represents a safe zone where your workout is effective but not overly strenuous to your body. By going beyond your training range, you risk hurting yourself. Consult Data Sheet 47.2 to determine and monitor your training range.

- Write a description of your exercise program to share with others, and be prepared to discuss it with the class.

- Plan to work out regularly with your partners. Try to arrange a schedule so that you can work out together. (Working out with friends usually makes it easier to maintain a fitness program.)

To Be Submitted

The description of your exercise program.

Your Training Range

Your training range (or target heart rate) gives you a safety zone while exercising. It is based on your age and heartbeats per minute. It is called a training range because it enables you to zero in on a level of exercise that is right for you.

Your Training Range

This example assumes an average fifteen-year-old student.

Always start at	220	beats per minute
Subtract your age	−15	
	205	beats per minute

Maximum safe heart rate = 205 beats per minute.

For a fifteen-year-old student, even one in good physical condition, going beyond 205 heartbeats per minute can be dangerous. The American College of Sports Medicine recommends that you calculate both 55% and 90% of your maximum safe heart rate to find the low and high end of your *training range.*

Multiply: $205 \times 0.55 = 112.75$. Round to 113 beats per minute. This is the low end of the range for a fifteen-year-old person. Exercising at this rate would result in a light workout. For people who have not exercised regularly during the past few months, exercising near the low end of the range is practical. As their conditioning improves, they can safely increase the level of activity and increase their heart rate.

Multiply: $205 \times 0.9 = 184.5$. Round to 185 beats per minute. This is the high end of the range. Exercising at this rate would result in a heavy workout.

The training range for an average fifteen-year-old student is between 113 and 185 heart beats per minute.

Your Training Range *(Cont'd.)*

Taking Your Pulse

You can keep track of your heart rate if you know how to take your pulse. Here is what to do:

1. Using your first two fingers (index and middle; no thumbs), press lightly on your carotid artery, which is located on the right side of your neck, straight down from the corner of your right eye. The artery is just under your chin. Gently put your fingers beneath your chin and feel for the pulse. You may need to move your fingers around a little, but the carotid is simple to find for most people.

2. Using a stopwatch or the second hand of a wristwatch, count the number of beats you feel for 10 seconds. This is your 10-second heart rate.

3. Since there are 60 seconds in a minute, multiply your 10-second heart rate by 6 to find your heart rate per minute.

Do not take anyone else's pulse, or let anyone else take yours. Keeping track of his or her pulse and heart rate is something each individual must do to avoid any confusion or mistakes.

Putting the Numbers Together

After finding your training range, take your pulse for 10-second periods several times throughout your workout. Multiply the number of beats times 6 to find your heart rate per minute. It should fall within your training range. If it is near the low end of the zone and you do not feel tired or out of breath, you can increase the level of your workout. If your heart rate per minute is near the top of your range, you should be careful not to exceed it. Of course, if you are becoming tired or having trouble catching your breath at any point during the workout, slow down.

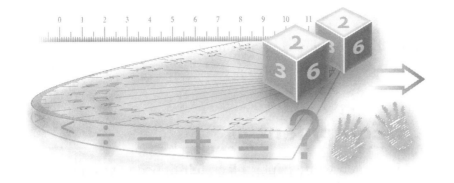

The big dance

School dances are popular events. While most students enjoy them, they do not think much about the work that goes into organizing and holding a big dance. They would be surprised at the amount of math (and money) that is involved.

Goal

Working in groups of three or four, students will plan a big dance for their school. They will consider such factors as entertainment, security, decorations, refreshments, and advertising, as well as raising the money to pay for the dance. They will write a summary of their plan and present their ideas to the class. *Suggested time:* Two class periods.

Math Skills to Highlight

1. Anticipating and estimating costs
2. Calculating costs
3. Determining the best buy

4. Applying the terms *revenue, costs,* and *profit*

5. Using technology in problem solving

Special Materials/Equipment

Calculators; $8\frac{1}{2}$- by 11-inch white paper; black felt-tipped pens and markers. *Optional:* Computers and printers; an overhead projector and transparencies for presentations; circulars from local grocery stores containing the prices of refreshments.

Development

Ask students how many of them like to attend dances. Probably most do. Explain that much work goes into organizing a successful dance. Although all the information students will need to complete this project is contained on Data Sheet 48.2, having students obtain prices for refreshments from the circulars of local grocery stores will broaden their choices. If you would like them to work from circulars, ask students to bring in grocery circulars from home a week or two prior to the project.

- Begin by explaining that students will work in groups of three or four to plan a dance for your school. Their group is responsible for obtaining a disk jockey (DJ) or another form of entertainment, arranging for security, providing refreshments, decorating, and creating advertising. To raise money to pay for the dance, they will need to set the price for tickets.

- Distribute copies of Student Guide 48.1, and review the information with your students.

- Explain the term *revenue.* Many students may not understand that it is the amount of money that, in this case, is obtained through the sale of tickets and refreshments. Also explain that *costs* refer to the money needed to pay for the dance, and *profit* is any money left over after the costs are subtracted from the revenue.

- Depending on your class, you may wish to review rounding and estimating.

- Hand out copies of Data Sheet 48.2, "The Costs of a Big Dance." Review the sheet with your students. Particularly point out that some assumptions are made on it. For example, groups are to assume that 300 students will attend the dance if a DJ plays music and provides entertainment, 250 will attend if a DJ simply plays music, and only 150 will attend if the music is supplied by CDs and managed by student volunteers. The difference in the number of students will affect the revenue. Also note that the PTA will help sponsor the dance by donating $250 toward its cost.

352

- If necessary, explain the difference between name brand and store brand items.

- If you have gathered grocery circulars, make them available to your students. Explain that the circulars give them a wider variety of items they may wish to select for refreshments and other items.

- Explain that most of the revenue to pay for the dance will come from selling tickets. To raise additional money, groups may decide to sell refreshments at a price above the cost at which they were bought. This will result in a profit that can be put toward the overall costs of the dance.

- Emphasize that groups should try to arrange what they think will be the best dance at the most reasonable cost to students.

- Distribute copies of Worksheet 48.3, "The Big Dance Costs/Revenue Tally Sheet," to the groups. The worksheet makes it easy to tabulate costs and revenues. Make additional sheets available.

- To publicize their dance, suggest that students design an advertisement. They may use markers or computers to create their ads.

- Remind students that each group is to write a summary of their plan for the big dance, including their anticipated costs and revenues. They are to designate a spokesperson to present their ideas to the class. Consider having students use an overhead projector in their presentations.

Wrap-Up

Each group shares its plan with the class. You might also display each group's written summary, advertisement, and final worksheet.

Extension

Create a committee to plan a real dance for your school. Be sure to check with your principal for your school's procedures in organizing a dance. Note that prices, depending on your locality, will vary from those presented in this project.

STUDENT GUIDE 48.1

The Big Dance

Situation/Problem

Your group has been selected to plan a big dance for your school. You will need to provide entertainment, security, refreshments, decorations, and advertising. Your school's PTA will help sponsor the dance through a $250 donation, but you will need to raise the rest of the money to pay for its costs through the sale of tickets and refreshments. Parent volunteers will chaperone the dance, sell tickets at the door, and sell the refreshments.

Possible Strategies

1. Discuss what factors make a dance successful. Is a disk jockey (DJ) important? Is it important for a DJ to play a wide variety of music? Is it important for a DJ to take an active role in the night and teach new dances, for example? How important are refreshments? How important is the cost of tickets? How important are decorations?

2. Consult Data Sheet 48.2 for prices associated with a big dance. Decide on the things you would like to have at your dance, round off prices, and calculate a rough estimate of your total costs. Subtract the $250 that the PTA will donate. This figure will represent how much revenue you will need to pay for your dance. Dividing this number by the total number of students you expect to attend the dance will give you the estimated cost of each ticket.

3. If the cost of your tickets is high, you might want to increase the prices of the refreshments you will sell. This will add to your overall revenue and help you to lower the prices of your tickets. You might instead decide to find a better price for your refreshments, use less expensive brands, or eliminate some refreshments. This will help you to lower your costs too.

The Big Dance *(Cont'd.)*

Special Considerations

- Decide what type of entertainment you will have. Note the differences in your expected turnout as listed on the data sheet. This will affect your ticket prices and revenue.

- Decide which refreshments, if any, you will provide. Estimate how much food you will need for refreshments. For example, if you decide to buy cans of soda, how many will you need? Also if you buy cans, you will need straws. If you decide to buy liter bottles, you will need cups in which to serve the soda. How many packs of snacks will you need? There are many decisions to make.

- Decide how much money you will need to pay for security.

- Decide how much you will spend on decorations. You may choose not to decorate to save money.

- Be willing to adjust your original decisions. You may find that costs prohibit you from having all the things you want at your dance. You may need to compromise.

- After you have decided on what you want at your dance, calculate your exact costs. Remember to add the PTA donation to your revenue. Use Worksheet 48.3 to tally your costs and balance them against your projected revenue. Note that for some items on the worksheet, the "Cost Each" will not apply.

- You must be able to pay for the dance entirely; you may not fall short. (To ensure that you have enough revenue to pay for the dance, you should anticipate that a few students you expect to come will not because of last-minute changes in plans. The prices of your tickets or refreshments should reflect this.) If you show a profit, the first $150 of any profit will be turned over to the PTA. Any amount over $150 will go into a general fund and will be used to defray costs for other school events.

- Design an advertisement on $8\frac{1}{2}$- by 11-inch white paper. Consider designing your advertisement on a computer. Be sure to include all necessary information: date, time, location, price, type of music, and if refreshments are available.

- Write a brief summary of your plan for the dance, and appoint a spokesperson to share your ideas with the class.

To Be Submitted

1. Summary
2. Advertisement
3. Worksheet

DATA SHEET 48.2

The Costs of a Big Dance

Entertainment: Disc Jockeys

"Mr. Smooth"—Music from the '80s to today. $525 for 3 hours; $150 for 1 hour extra.

"Wild Man"—Your choice of music from the '60s to today. Provides entertainment and teaches new dances. $900 for three hours; $300 for 1 hour extra.

An alternative: Student volunteers play CDs for no cost.

Notes on expected attendance: With Mr. Smooth, you can expect 250 students to attend the dance. With Wild Man, you can expect 300. CDs played by student volunteers will result in 150 students attending the dance.

Security

You will need at least one off-duty police officer for 150 students, and another officer for every 100 additional students after that. Cost: $40 per hour for each officer.

Refreshments

Soda—brand name:	2-liter bottles, $1.09 each
	12-ounce cans, $2.99 per six-pack
store brand:	2-liter bottles, $0.89 each
	12-ounce cans, $1.79 per six-pack

Snacks—small, individual packs of potato chips, pretzels, corn chips, and cheese chips: packs of 6 1-ounce bags, $2.99

Spring water—16.9-ounce bottles, $4.99 for 24

Additional items—

Straws, 250 for $1.29

Plastic cups, 100 (5-ounce) for $2.69

Napkins, 120 for $1.89

Decorations

A minimal amount of decorations, $45

More elaborate decorations, including streamers, some balloons, and wall displays, $75

WORKSHEET 48.3

The Big Dance Costs/Revenue Tally Sheet

COSTS

Item	Cost Each	Total
	Total Costs	

REVENUE

Source	Total
Total Revenue	

Subtract *Total Costs* from *Total Revenue*. The difference should either be zero or show that you have a profit.

Total Revenue − Total Costs = _____

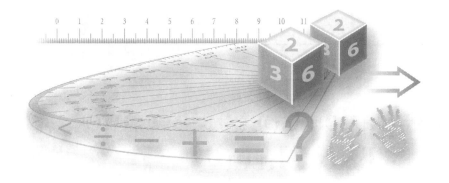

The numbers game

Many students enjoy solving math puzzles and games. In this project, students work with common phrases that are related to numbers. To add excitement to the challenge, you may wish to award prizes to the students who achieve the highest score.

Goal

Working in pairs or groups of three, students are to find the meaning of each "number" phrase on a list they are given. They will also be encouraged to compile a list of number phrases of their own. *Suggested time:* One class period (a partial period to introduce the project and a partial period to discuss answers). Much of the work for this project will be done out of class.

Math Skills to Highlight

1. An awareness of how numbers are a part of our lives other than the obvious tasks of computation and analysis
2. Using technology in problem solving

Special Materials/Equipment

Dictionaries; writer's stylebooks; math reference books. *Optional:* Computers with Internet access for research.

Development

Pose some number phrases such as the "five senses" (hearing, sight, smell, taste, and touch), the "fourth dimension" (time), and "one-way ticket" (a ticket that takes you to a destination but not back again). Discuss what they mean, and explain that number phrases are common in our language. Ask your students to offer some they know.

- Begin the project by explaining to your students that they will work in pairs or groups of three. They will be given a list of number phrases and will have to define what they mean.

- Distribute copies of Student Guide 49.1, and review the information with your students. Point out the scoring strategy for this project. Groups will earn 2 points for each number phrase they define correctly and 3 points for each new phrase that they can think of that is not on the list.

- If you decide to make this project a competition, point out that the winners will likely be decided by the number of additional phrases they find. Note that any additional phrases must be verified in reference sources. Students should also provide a key for the meanings of their phrases. Consider giving a prize—perhaps a homework pass—to the winners.

- Hand out copies of Worksheet 49.2, "Number Phrases," and review it with your students. The sheet contains twenty-five phrases based on numbers. Following is the answer key:

  1. *One-liner*—a short joke.
  2. *In two shakes of a lamb's tail*—quickly. (Have you ever seen a lamb shake its tail?)
  3. *Three-dog night*—a very cold night. (On cold nights in the Arctic, three sled dogs are needed to keep warm when sleeping.)
  4. *On all fours*—on hands and knees.
  5. *Take five*—to take a break. (A five-minute break.)
  6. *Six of one and a half-dozen of the other*—equally accepted.
  7. *Seven seas*—the seven oceans. (North and South Pacific, North and South Atlantic, Arctic, Antarctic, and Indian. Granted, this stretches modern geography a bit, but that is how the phrase originated.)
  8. *Eight-hour day*—the typical workday, 9:00 A.M. to 5:00 P.M.

9. *On cloud nine*—very happy. (Originated from the number nine, which was considered to be a perfect number.)

10. *Top 10*—the first ten of a list. (Originally for pop music.)

11. *Eleventh hour*—the latest possible time.

12. *Twelvemonth*—a year.

13. *Catch-22*—a situation from which there is no solution. (Taken from the title of the Joseph Heller novel, *Catch-22*.)

14. *Twenty-three skidoo*—go away. (Refers to Twenty-Third Street in New York City during the 1920s and 1930s. It was from that street that many railroads left the city.)

15. *Twenty-four-hour*—a period lasting day and night.

16. *Forty winks*—a short nap. (Thought to be an arbitrary number.)

17. *Fifty-fifty*—equal.

18. *Eighty-eight*—a piano. (A standard piano has eighty-eight keys.)

19. *Ninety-nine times out of a hundred*—often.

20. *Hundred and one*—many.

21. *One hundred percent*—entirely, completely.

22. *Thousand and one*—very many.

23. *Sixty-four-thousand-dollar question*—the most important question. (Based on the 1940s radio quiz show, which featured the $64 question. The $64,000 question is the updated version.)

24. *Feel like a million*—to feel terrific.

25. *A million to one*—very low chance for success.

Discuss where students might find the meanings to the phrases. Some possibilities are dictionaries, writer's stylebooks, and math reference books, as well as online sources. Suggest that groups meet in the library during free periods or after school if necessary. Students might also consider asking their parents or grandparents for help. Older relatives might be familiar with phrases such as *twenty-three skidoo*. Getting the family involved with this project can make it an enjoyable activity for everyone.

Caution teams not to give answers to each other.

Wrap-Up

Provide the answers, and ask students to tally their scores.

Extension

Compile the additional phrases that students found into a new list. Make copies of this list and distribute it to students who are interested in continuing the project.

STUDENT GUIDE 49.1

The Numbers Game

"Cloud 9"

Situation/Problem

You and your partner or group are to find the meanings of the number phrases listed on Worksheet 49.2. Also try to write down other number phrases. Each correct answer is worth 2 points, and each additional phrase you provide is worth 3 points.

Possible Strategies

1. Skim the list, and write down the meanings of any phrases you know.
2. Write down any other number phrases (not on the list) that you know. To receive 3 points for each phrase, you must be able to verify it and its meaning through reference sources.

Special Considerations

- Consult dictionaries, writer's stylebooks, and math reference books for the meanings of number phrases. You might also consult online sources.
- Ask your parents and grandparents if they know the meanings of any of the phrases.
- Correct your answers, and compute your score.

To Be Submitted

Completed Worksheet 49.2.

WORKSHEET 49.2

Number Phrases

Write the meaning of each phrase given. Each correct answer is worth 2 points. At the end of this worksheet, write any other number phrases you know and their definition. These are worth 3 points each, but you must be able to verify the accuracy of your phrases through reference sources. Write the source after each phrase.

1. One-liner

   _____

2. In two shakes of a lamb's tail

   _____

3. Three-dog night

   _____

4. On all fours

   _____

5. Take five

   _____

6. Six of one and a half-dozen of the other

   _____

7. Seven seas

   _____

8. Eight-hour day

   _____

9. On cloud nine

   _____

10. Top 10

   _____

Number Phrases *(Cont'd.)*

11. Eleventh hour

   _____

12. Twelvemonth

   _____

13. Catch-22

   _____

14. Twenty-three skidoo

   _____

15. Twenty-four-hour

   _____

16. Forty winks

   _____

17. Fifty-fifty

   _____

18. Eighty-eight

   _____

19. Ninety-nine times out of a hundred

   _____

20. Hundred and one

   _____

Number Phrases *(Cont'd.)*

21. One hundred percent

_____

22. Thousand and one

_____

23. Sixty-four-thousand-dollar question

_____

24. Feel like a million

_____

25. A million to one

_____

Other Number Phrases and Their Definitions

_____

_____

_____

_____

_____

_____

_____

_____

_____

_____

_____

Planning a sundae party

Students like to have parties. Holiday parties, end-of-the-year parties, good-bye parties, or a party to reward students for their achievements are common in many school systems. The parties are usually planned by teachers or parents, or both. In this project, students will plan an ice cream sundae party for the class. You may wish to celebrate the completion of this project with a real sundae party, based on their suggestions.

Goal

Students will work in groups of three or four to plan a sundae party for their class. At the end of the project, a spokesperson for each group will share their plan for the party with the class. *Suggested time:* Two class periods.

Math Skills to Highlight

1. Using estimates to determine serving size
2. Rounding numbers
3. Estimating the total costs of a sundae party
4. Calculating the total costs
5. Using technology in problem solving

Special Materials/Equipment

Calculators. *Optional:* Grocery circulars containing the prices of ice cream and toppings; an overhead projector and transparencies for presentations.

Development

Although Data Sheet 50.2 contains the prices for ice cream, toppings, and other materials needed for a sundae party, you may wish to collect grocery circulars to make the project more realistic for your students. A week or two prior to the project, ask students to bring in circulars from home. *Note:* Before distributing the circulars to your class, check them to make sure they include the prices for ice cream and toppings. Grocery circulars do not always contain these items.

- Begin the project by explaining to your students that they will be working in groups of three or four to plan a sundae party for the class. They will be required to choose the type and amount of ice cream, any toppings they wish, and any other materials they will need, such as napkins, plastic cups, and spoons. Since students will need to pay for the party by contributing an equal sum per student, groups should attempt to keep the costs of the party within reason.

- Hand out copies of Student Guide 50.1, and review the information with your students. Emphasize that groups are to plan the best party for the most reasonable cost.

- Distribute copies of Data Sheet 50.2, "Prices for a Sundae Party." Review the prices of the various items with your class. In particular, point out the serving sizes on the bottom of the Sheet. Students will need to know these to estimate how much ice cream and toppings to buy. Also point out the wide variety of toppings, and tell the groups that they do not have to provide all the toppings for the party. They may choose the ones they want. If necessary, explain the difference between name brand and store brand items.

- If you are making grocery circulars available, tell students that they may use these as well to make their choices. Also, if the circulars contain the prices for fat-free ice cream, frozen yogurt, sherbet, or sorbet, you may wish to broaden the project and permit groups to choose these items in addition to those on the data sheet.

- If necessary, review rounding for estimates.

- Encourage students to estimate the amounts and costs of the items they will need for their party, find out the estimated cost per student, and then work from there to make their selections.

- Hand out copies of Worksheet 50.3, "Tallying the Costs of a Sundae Party," for students to use in calculating their costs.

- Remind students that they should appoint a spokesperson to describe their party choices to the class. Suggest that they use an overhead projector to present their conclusions. They can use Worksheet 50.3 to make a transparency to highlight their results.

Wrap-Up

Students share their sundae party plans with the class.

Extension

Have a sundae party. You might have students decide on the best plan proposed by the different groups, or simply list the items students want on the board or an overhead projector and together calculate the costs.

If you have the party, here are some tips:

- Keep things practical.

  Limit the choices of ice cream. If you provide more than three, some students will have trouble deciding which flavors they want. That slows down the serving process.

  Avoid ice cream containers that have two or three flavors. It is better to have only one flavor per container. This eliminates the problem of students trying to spoon out only the flavor they want.

  Use whipped topping that comes in a container and must be spooned out. Avoid whipped cream in a spray can. Some students might be tempted to spray it on each other rather than on their ice cream.

- Collect all the money a few days in advance.

- Enlist parent volunteers to buy the ice cream and bring it to class at the time for the party. This eliminates the need to store the ice cream at school.

Planning a sundae party

- If you must buy the ice cream yourself, make sure you have enough freezer space at home and at school to store it.

- If you have tables, lay plastic tablecloths over them to reduce the mess and cleanup. (Garbage bags are an inexpensive option, especially if the school janitor is willing to give you some.) If you do not have tables, you may simply push desks together and cover them. Better yet, see if you can have the party in the cafeteria.

- Have handy plenty of napkins, paper towels, a sponge, and a small bucket filled with water.

- Line the classroom wastebasket with a garbage bag. This will prevent the leakage of any melting ice cream.

- Set the ice cream, toppings, dishes, scoops, spoons, napkins, and anything else out on your tables, and let your students serve themselves buffet style. Have separate scoops for each container of ice cream and spoons for each topping.

Enjoy! It will be a great party.

STUDENT GUIDE 50.1

Planning a Sundae Party

Situation/Problem

Your group will plan a sundae party for the class. Keeping in mind that class members will have to pay for the party, your task is to provide a great party with different flavors of ice cream and a variety of toppings at a reasonable cost. At the end of the project, your group's spokesperson will describe your plan for the party to the class.

Possible Strategies

1. Discuss the types of ice cream and toppings your group feels would be popular at a sundae party.

2. Decide what you feel is a fair price for individual students to pay for the party.

3. Make a rough estimate of your total costs. See Data Sheet 50.2 for items and prices. Select the items and amounts you would like for your party, round off the prices, and estimate your costs. Dividing by the number of students who wish to participate in the party will give you an estimate of the cost per student. If the price is higher than what you originally decided was a fair price, you will need to eliminate some items or discuss raising the price each student must pay. If your estimate is less than what you feel is a fair price, you may decide to add some things.

Special Considerations

- Be realistic in your estimates of what students will pay for a sundae party.

- Be willing to compromise over the items you select. To keep the price of the party reasonable, you may need to eliminate some things you would like.

- Pay close attention to serving sizes when you estimate how much ice cream or toppings to buy. (Unless you use giant scoops, two scoops of ice cream equal about 4 ounces, a typical serving size.) If you want students to have more than the typical serving size, you will need to buy more ice cream. Also, if students will be serving themselves, they might be more generous with the amount of ice cream they scoop out than you planned on. You may want to plan to buy a little extra to make sure that you have enough. If you have any ice cream left over, you can always offer seconds.

369

Planning a Sundae Party *(Cont'd.)*

- You must buy sufficient materials for cleanup.
- As you select the items you want, maintain a running tally of your costs. Use Worksheet 50.3.
- Appoint a spokesperson to share your party plan with the class.

To Be Submitted

Your worksheet.

Notes

_____

_____

_____

_____

_____

_____

_____

_____

_____

_____

_____

_____

_____

_____

Name _____

Prices for a Sundae Party

Following are the costs of various items you will need for a class sundae party.

Ice Cream and Serving Items

Ice cream: $\frac{1}{2}$ gallon: brand name, $4.89; store brand, $2.49; 1 gallon: brand name, $7.99; store brand, $4.89

Plastic bowls, 12 ounces, 50 for $1.89

Plastic spoons, 24 for $2.19

Napkins, 100 for $1.89

Serving spoons for toppings, 2 for $1.00

Ice cream scoops, $2.69 each

Toppings

Whipped topping, 8 ounces, $1.49; 16 ounces, $1.99

Chocolate syrup, 24 ounces, $1.79

Strawberries, 20 ounces, $2.29

Maraschino cherries, 25 for $1.39; 50 for $2.09

Chocolate chips, 12 ounces, $2.69; 24 ounces, $5.29

Sprinkles, 3 ounces, $1.69; 9 ounces, $2.99

Cleanup

Plastic tablecloth, 54″ by 108″, $1.99 each

Paper towels, 64 sheets for $1.69 each

Sponge, 5″ by 8″, $1.19 each

Small bucket, 11 quarts, $6.49 each

Tips

Each half-gallon of ice cream contains 16 4-ounce servings. (A 4-ounce serving equals one-half cup.)

Each gallon of ice cream contains 32 4-ounce servings.

Serving sizes for toppings listed above:

> Whipped topping—8 ounces, 28 2-tablespoon servings; 16 ounces, 56 2-tbsp. servings
>
> Chocolate syrup—24 ounces, 17 2-tablespoon servings
>
> Strawberries—20 ounces, 15 2-tablespoon servings
>
> Chocolate chips—12 ounces, 12 2-tablespoon servings; 24 ounces, 24 2-tbsp. servings
>
> Sprinkles—3 ounces, 4 2-tablespoon servings; 9 ounces, 12 2-tbsp. servings

WORKSHEET 50.3

Tallying the Costs of a Sundae Party

Item	Number of Each	Cost Each	Total

Total Cost _____

Divide the total cost by the number of students participating in the party to find the cost per student:

Total Cost ÷ Number of Students = _____ Cost per Student

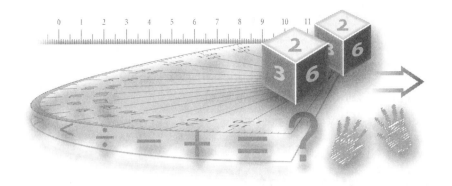

Going on vacation

Many, if not most, of your students have gone on vacations, but not very many have taken part in planning one. This project gives them the chance to plan a vacation in which they imagine they drive to a major resort.

Goal

Working in groups of three or four, students will plan a vacation to a well-known resort. Given a specific budget, they will need to calculate their round-trip costs for travel, lodging, food, and entertainment. At the end of the project, a spokesperson from each group will share the group's vacation plan with the class. *Suggested time:* Two to three class periods.

Math Skills to Highlight

1. Measuring distance
2. Calculating miles per gallon
3. Making decisions regarding travel options

4. Rounding numbers

5. Estimating costs for travel, lodging, and visiting a major resort

6. Calculating the overall costs for a vacation

7. Using technology in problem solving

Special Materials/Equipment

Calculators; reference books or atlases to estimate distances. *Optional:* Travel or vacation brochures that provide information about specific resorts; computers with Internet access for research; an overhead projector and transparencies for presentations.

Development

You have several alternatives in developing this project. You might wish to focus your students on a specific resort, for example, Walt Disney World in Florida, or give them a choice of resorts (Walt Disney World, Disneyland in California, or Busch Gardens/Williamsburg in Virginia), or let them select their own. Giving students several choices will add variety to the project; however, the focus of the project is on resorts that contain theme or amusement parks.

You might also consider sending away for brochures about specific resorts or check the Web sites of resorts for information. This can help make the project realistic. If you decide to send for information, be sure to request brochures well in advance, perhaps six to eight weeks ahead of time.

- Begin this project by explaining to your students that they will be working in groups of three or four to plan a vacation to a major resort. They are to imagine that their group has won a nationwide contest for the group "Most Likely to Succeed in Math," and the prize is $4,500, which they are to spend on a vacation.

- Tell them the places from which they may choose. They are to plan a round-trip by car. They will assume that they will be traveling with a parent chaperone who will handle the driving; the group will pay for this parent's expenses. (A parent who is driving three or four kids on a vacation deserves to go for free!)

- Hand out copies of Student Guide 51.1, and review the information with your students. Emphasize that they are to consider all the costs associated with their vacation except personal items such as souvenirs. They may stay as long as they can afford to, but they must keep within their budget. This will depend on the distance they must travel as well as the lodging, food, and admissions plans they select.

- Hand out copies of Data Sheet 51.2, "The Costs of a Great Vacation." Review the material closely with your students. Some will likely have questions. Note that they have plenty of options.
- If you acquired travel or vacation brochures you would like your students to use in addition to the data sheet, hand them out and provide any necessary instructions. You might also suggest that students visit the Web sites of destinations.
- Distribute copies of Worksheet 51.3, "Calculating Vacation Costs." Students should use this to tally their expenses. Make extra copies available.
- If necessary, review rounding and estimation skills with your students.
- Depending on the destinations your students select, you may wish to schedule a class period (or at least a partial period) in the library so that students may consult atlases for estimates of distances. Students may also have maps in their social studies text they can use. They can also check distances online.
- Remind groups to appoint a spokesperson to share their vacation destination and plan with the class. Suggest that students use an overhead projector to present their conclusions to the class. They can use Worksheet 51.3 to make a transparency to highlight their results.

Wrap-Up

Spokespersons share their group's plan.

Extension

Suggest that students explore plane and train travel costs to their destination. After considering the costs, which is least expensive: Car, plane, or train? Which is most practical?

STUDENT GUIDE 51.1

Going on Vacation

Situation/Problem

Imagine that your group has won the coveted "Most Likely to Succeed in Math" award. Your prize is $4,500, which is to be applied to a vacation. Your teacher will provide you with the places from which you can choose to go. You will be traveling by car (one of your parents will drive and act as chaperone), and you must plan for all the expenses of your trip. You may not exceed your $4,500 prize. On completion of the project, a spokesperson will share your vacation plan with the class.

Possible Strategies

1. If you have a choice of places, decide which one you will visit.

2. Estimate the distance you must travel, and select the type of vehicle in which you will drive. Consult Data Sheet 51.2. A compact car gets more miles to the gallon, but you will feel cramped on a long trip. You will be more comfortable in a minivan or SUV, but minivans and SUVs gobble gas. Based on the distance and gas mileage your vehicle is capable of, estimate your fuel costs. Take your total distance (round-trip), and divide by miles per gallon. Multiply the answer by the cost of a gallon of gasoline. Your answer will be your total costs for gas.

3. Assuming you will be able to travel 500 miles per day (about 60 mph for 8 hours), estimate how long it will take you to arrive at your destination. For example, a 1,000-mile trip would require two days of driving, and you would need to stay one night in a motel.

4. Using Data Sheet 51.2, plan your vacation. Include all lodging, food, and ticket prices, and make a rough estimate of your costs. Add to these the cost of gasoline. This should give you an estimate of your overall costs for your entire vacation. If your expenses surpass the $4,500 you have available, you must reduce your costs. If your expenses are below your $4,500, you can add some things.

Going on Vacation *(Cont'd.)*

Special Considerations

- Do not include costs for souvenirs or extras. Individuals are responsible for these expenses.
- Remember that travel and lodging costs must be calculated for a round-trip.
- If you have boys and girls in your group, you will need two motel or hotel rooms, one for the boys and one for the girls. This will add to your costs. (Groups of all boys or all girls may stay in one room.)
- You have various choices and options. Study the data sheet carefully, and select those that you feel will result in an outstanding vacation.
- You may remain on vacation as long as you can pay your expenses.
- You must pay all of the expenses of your parent chaperone, including lodging (if he or she stays in a separate room), food, and any admission prices.
- Use Worksheet 51.3 to calculate your costs. Be sure to write down your total costs. For example, if you decide to stay at a resort's superior hotel for four days, one room will cost $1,400.
- Appoint a spokesperson to share your plan with the class.

To Be Submitted

Your worksheet

Notes

_____

_____

_____

_____

_____

_____

_____

_____

The Costs of a Great Vacation

You can expect the following costs for your vacation. The prices are general, and some assumptions have been made.

Travel Costs

Gas (the current cost per gallon):

A small car gets 45 mpg (miles per gallon)

A family sedan gets 25 mpg

A minivan or SUV gets 20 mpg or less

Meals:

$25 per day per person (mostly fast food)

$45 per day per person (a good breakfast, fast food for lunch, a full meal for dinner)

Lodging (up to five people per room):

Superior motel, $99 a room per night (indoor and outdoor pools, restaurant, cable TV, free movies, workout room)

Good motel, $69 a room per night (outdoor pool, cable TV, restaurant nearby)

Economy motel, $39 a room per night (cable TV)

At the Resort

Lodging (up to five people per room):

Superior hotel, $350 a room per night (huge pool, four restaurants, twenty-four-hour snack shop, health club open to guests, shopping boutiques, arcade)

Good hotel, $225 a room per night (midsized pool, restaurant, workout room)

Economy hotel, $150 a room per night (small pool, restaurant; ten-minute drive to the resort)

Admission to attractions:

One-day pass, $45; two-day pass, $75; four-day pass, $140. With a four-day pass, a fifth day pass may be purchased for $30.

Meal Plans:

$50 per person each day of stay (includes lunch and dinner at selected restaurants throughout the resort).

$65 per person each day of stay (includes breakfast, lunch, and dinner at all the restaurants throughout the resort).

Name _____

Calculating Vacation Costs

Destination: _____

Number of Miles (Round-Trip): _____

Number of People in Group: _____

Include total costs for each item.

Travel Costs (Round-Trip)

Gasoline: miles of trip/mpg × price of gas

_____/_____ × _____ = $ _____

Lodging: number of nights at motel × cost per night × number of rooms

_____ × _____ × _____ = $ _____

Meals: number of days × cost per day × number of persons

_____ × _____ × _____ = $ _____

Subtotal $ _____

Resort Costs

Hotel: number of nights × cost per night × number of rooms

_____ × _____ × _____ = $ _____

Admissions: cost × number of persons

_____ × _____ = $ _____

Meals: number of days × cost per day × number of persons

_____ × _____ × _____ = $ _____

Subtotal $ _____

Adding both subtotals together will give you the total cost of your vacation.

_____ + _____ = _____

Subtotal   **Subtotal**   **Total Cost**

Math and Life Skills

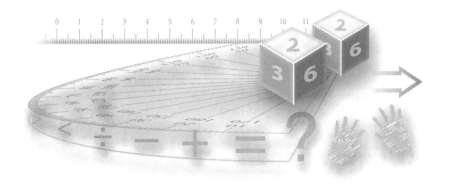

Making a budget

Many students feel that money is in endless supply. They buy things they want without considering whether they really need them. Part of the problem is that they have no way of keeping track of how they spend their money. For some students, a budget can be helpful. Creating a budget will not guarantee that money will be spent wisely, but at least it enables individuals to monitor their income and expenses.

Goal

Working individually, students will create a monthly budget to manage their income and expenses. *Suggested time:* Two class periods.

Math Skills to Highlight

1. Making a budget
2. Estimating income and expenses

3. Applying the terms *income, expenses,* and *surplus*

4. Making wise choices when spending money

5. Using technology in problem solving

Special Materials/Equipment

Calculators.

Development

Discuss the importance of having a budget and how it is useful in managing personal finances. Budgets enable people to see how and where they spend money, helping them to adjust spending habits so that they can meet their financial goals. You might also discuss the importance of savings, perhaps for a car, college, or emergencies, and point out that budgets can help people manage their money so that they have some left over to save.

- Begin the project by explaining that students will work individually to create a monthly budget that monitors their income and expenses. If you have students who prefer not to reveal their finances, suggest that they pick fictional, though realistic, amounts to work with. By offering this option, you relieve students of any pressure they might otherwise feel regarding their personal situation with money. They will nevertheless benefit from the project.

- Distribute copies of Student Guide 52.1, and review the information with your students.

- Distribute copies of Data Sheet 52.2, "Tips for Making a Budget." Review the material with your students, and point out that the worksheet offers some important budget terms and a step-by-step guide for making a budget.

- Discuss the terms *income, expenses,* and *surplus.*

- If necessary, review estimation with your class.

- Although some students might be inclined to underestimate their expenses in order to stay within their budget, caution them that this will only lead to a shortage of funds. Rather, they should overestimate expenses. It is always nice to have a surplus.

- Hand out Worksheet 52.3, "Working on Your Budget." Students should use this worksheet for their budget. Remind them to be sure to list all income and expenses.

Wrap-Up

Collect the worksheets, and discuss any problems or insights students might have had in making their budgets.

Extensions

Invite a financial planner to class to discuss the importance of budgets and planning for the future. You might also wish to mention that various software programs are designed to help people create budgets and manage their finances. Perhaps you or some of your students are familiar with these programs. You may wish to discuss them.

STUDENT GUIDE 52.1

Making a Budget

Situation/Problem

You are to make a monthly budget that will help you to keep track of your income and expenses.

Possible Strategies

1. Estimate your income by adding up all the money you receive during the month.
2. Make a list of all the things you spend money on.
3. Estimate your expenses by adding up the costs of all the things you buy.
4. Subtract your expenses from your income.

Making a Budget *(Cont'd.)*

Special Considerations

- Carefully review Data Sheet 52.2. It provides detailed information about income and expenses.
- If your expenses are more than your income, you are spending too much and must reduce your spending or increase your income. If you have money left over, you have a surplus. You may either save it or buy some other things you would like.
- Use Worksheet 52.3 to make your budget.

To Be Submitted

Your worksheet.

Notes

_____

_____

_____

_____

_____

_____

_____

_____

_____

_____

_____

_____

_____

DATA SHEET 52.2

Tips for Making a Budget

A budget is an excellent tool for managing money. A good budget contains your sources of income and your expenses. Budgets may be weekly, monthly, or even yearly. Use Worksheet 52.3 to make a monthly budget based on these guidelines:

1. List all sources of your income for the month and the amount of money you receive from each (for example, an allowance, part-time job, and errands). If a source of income varies, such as a gift, or you are not sure about it, underestimate its amount.

2. Add up your income.

3. List all of your expenses and the amount you pay for each. Expenses include spending money on food (lunch and snacks), clothes, entertainment (movies, DVDs, CDs), sports events and equipment, and car costs. Many people also include savings as an expense. By doing this, they are more likely to place money in a savings plan each month.

4. Add up all your expenses.

5. Subtract your total expenses from your total income.

    If the difference is zero, your budget is balanced.

    If you have money left over, you have a surplus and are to be congratulated. You may put more money in savings, keep it handy for unexpected expenses, or spend it.

    If your expenses are greater than your income, you must adjust your budget by spending less or increasing your income.

Working on Your Budget

List your monthly sources of income and their amounts. List your monthly expenses and the amounts. Find the totals; then subtract your expenses from your income.

Sources of Income	Amount	Expenses	Amount

Total: $ _____   Total: $ _____

$ _____ − $ _____ = $ _____
    Income   Expenses

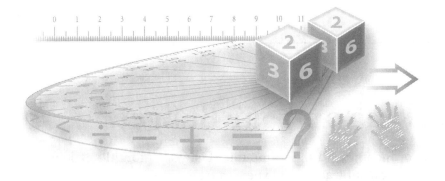

A floor plan of my room

The rooms of teenagers are often a mess. Sometimes this is because items are simply thrown all about; in other cases, the furniture in the room is not arranged in a manner that makes best use of the space. A little rearranging can help to make a room more comfortable, keep clutter under control, and make the room appear less messy. Rearranging furniture can be a big job, though, especially if the plan is to pull beds, bureaus, desks, and chairs around to see how they fit best. A better way is to create a floor plan of the room and try various arrangements on paper before moving anything.

Goal

Working individually, students will make a floor plan of their room. They will need to measure their room and the length and width of the furniture they have. They will then create a scale drawing of their room, make scale drawings of their furniture, and determine if the furniture in their room can be arranged more effectively. Of course, they may find that the best arrangement is the current one. *Suggested time:* Two class periods. The measuring will be done at home.

Math Skills to Highlight

1. Taking measurements
2. Making a scale drawing
3. Using technology in problem solving

Special Materials/Equipment

Tape measures or yardsticks; rulers; scissors. *Optional:* Calculators.

Development

Ask students to think of their room. Is it messy? Cluttered? Now ask them to think about how their furniture is arranged, and explain that sometimes rearranging the furniture in a room can make better use of space and at the same time make the room more attractive or visually appealing.

- Begin this project by explaining that students will work individually to create a floor plan of their room that makes the most effective use of the room's space.

- Distribute copies of Student Guide 53.1, and review the information with your students. Make certain that students understand how they are to create a floor plan, particularly finding the dimensions of their room and selecting a scale.

- Hand out copies of Data Sheet 53.2, "How to Measure a Room," and discuss the information. Suggest that all the measurements be made in inches; using inches instead of feet for the measurements will make it easier to set up a scale.

- Encourage students to draw a sketch of their room first and include measurements and the approximate locations of doors and windows. Not only will this help them to visualize the overall room, but they can also use the sketch to record their room's dimensions. Later they can take the dimensions from the sketch and transfer them to their scale drawing.

- Explain that students may use either tape measures or yardsticks to measure the dimensions of their room. Perhaps their parents have these supplies, or you may ask your school's home arts, art, or science teacher if he or she can lend you some tape measures.

- Instruct students to measure the length and width of their furniture, and record the measurements on the same sheet as the sketch.

- After they have taken all the measurements and recorded the measurements on the sketch, students should determine a scale. Remind students to consult their student guide for instructions on how to determine a scale.

- Hand out two copies of Worksheet 53.3, "A Grid for a Floor Plan," to each student. (One is for a scale drawing, and the other is for drawing the room's furniture.) Have extra copies available.

- Instruct students to make a scale drawing of their room on the worksheet, including all the measurements they recorded on their sketches. Remind students to include the scale on their drawings.

- On another copy of the worksheet, students are to draw their furniture according to the scale they chose. They should then cut out the figures and place them on the scale drawing of their room. They should try various arrangements of the furniture to see which uses the space of their room the best.

- After finding what they feel is the best arrangement of their furniture, students are to draw their pieces of furniture in place on the scale drawing. (If students find that a new arrangement of furniture uses the space of their room better than the existing one, they may wish to ask their parents if they can reorganize their room.)

- Students should be prepared to discuss any changes they would make in their current floor plan or explain why the current floor plan is best.

Wrap-Up

Display the floor plans.

Extension

Students may wish to make scale drawings of other rooms in their home.

STUDENT GUIDE 53.1

A Floor Plan of My Room

Situation/Problem

You will make a floor plan of your room; it will include the lengths of the walls, location of the windows and doors, and the arrangement of furniture. Your floor plan may be of your current room, or an example of how you might rearrange your furniture to make better use of your room's space.

Possible Strategies

1. Make a rough sketch of your room.
2. Measure all the walls of your room accurately.
3. Measure all doors and windows.
4. Measure the length and width of all pieces of furniture.
5. Record all measurements on your sketch.

A Floor Plan of My Room *(Cont'd.)*

Special Considerations

- Consult Data Sheet 53.2 for guidelines on how to measure your room.

- Always measure in inches, and double-check all measurements.

- Choose a scale. Consider using $\frac{1}{4}$ inch = 12 inches or $\frac{1}{2}$ inch = 12 inches. Your scale will depend on the overall size of your room. Choose the scale that works best. To create a scale, follow these suggestions:

  Note that the side of each square of the grid on Worksheet 53.3 is $\frac{1}{4}$ inch long.

  If the longer length of your room is greater than 180 inches (15 feet), use the scale $\frac{1}{4}$ inch = 12 inches. If the shorter length is greater than 144 inches (12 feet), use the scale $\frac{1}{4}$ inch = 12 inches. If neither of these is true, use the scale $\frac{1}{2}$ inch = 12 inches.

  If you use the scale of $\frac{1}{2}$ inch = 12 inches, remember that every side of two squares on the grid represents 12 inches of your room. A wall 120 inches long would be equal to 20 squares.

  If you use the scale of $\frac{1}{4}$ inch = 12 inches, remember that every side of the squares on the grid represents 12 inches of your room. A wall 216 inches long would be equal to 18 squares.

- Make a scale drawing of your room on Worksheet 53.3. Label all walls, windows, and doors. Be sure to indicate which way the door swings with a line that represents the width of the door. Your floor plan is now complete.

- Using another copy of Worksheet 53.3, make a scale drawing of each piece of furniture. Use the same scale as you chose for your room. Write the name of each piece of furniture on it so that you do not confuse it with other pieces.

- Cut out each piece of furniture, and set it on your floor plan. Arrange the pieces in various ways. You may find a better way to organize the furniture in your room, or you may find that the arrangement you now have is best.

- When you are satisfied you have found the best arrangement, draw the pieces of furniture on your floor plan. You may wish to label your floor plan as "Current" or "Proposed" Arrangement of Furniture.

- Be prepared to discuss your floor plan with the class and why you feel this one uses your room's space most effectively.

To Be Submitted

The worksheet that contains the final copy of your floor plan.

DATA SHEET 53.2

How to Measure a Room

The following suggestions will help you make accurate measurements of your room.

1. Imagine that you are looking down on your room from the ceiling. Make a rough sketch of the floor of your room, and include the approximate positions of the furniture: bed, desk, chairs, and so on. This will help you to visualize its contents in their approximate locations.

2. Use a tape measure or yardstick to measure your room. Use inches for all measurements.

3. Measure the length of each wall. Start at the beginning of the left corner and measure to the right corner. Record your measurements on your sketch.

4. To position doors and windows accurately on your floor plan, you must first find their measurements:

    Start at the left corner of the wall on which the door or window is located, and measure to the left edge of the door or window. Record the distance on your sketch.

    Measure across the opening of the door or window from edge to edge. Record this distance on your sketch.

    Measure from the right edge to the right corner of the wall, and record the distance on your sketch.

    The sum of these measurements should equal the distance of the entire wall.

5. Mark the position of doors and windows on your sketch, and include measurements. (Remember that you are looking down on your room from above. You cannot draw a full window or door. Indicate them by narrow rectangles or double lines.)

6. Show the way any doors swing open with a line that represents the width of the door.

Name _____

A Grid for a Floor Plan

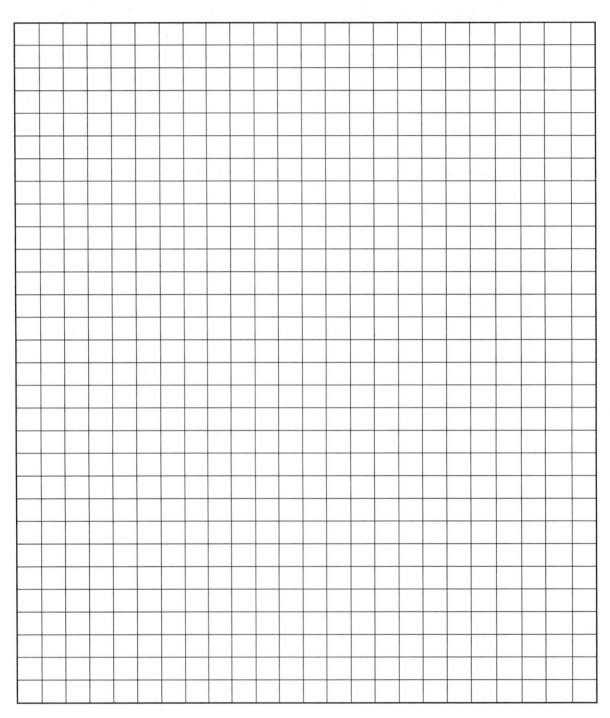

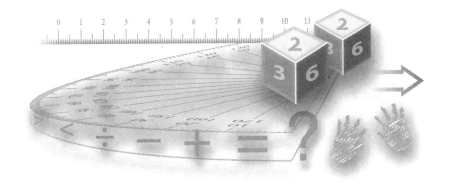

The costs of pets

Many of your students probably have pets, from dogs and cats to turtles and snakes. Pets offer companionship, enjoyment, and satisfaction. Most also come with hefty costs. While students may love their pets and even help take care of them, few think about the expense of having a pet.

Goal

Working in pairs or groups of three, students are to imagine that they are interested in buying a pet. Before they do, they are to research the expenses related to that pet and, along with its purchase price, determine potential costs for veterinarian services, food, cages, toys, and other expenses for a one-year period. Students are to write a summary report of their findings and share their findings orally with the class. *Suggested time:* Two to three class periods.

Math Skills to Highlight

1. Researching various cost factors
2. Estimating costs
3. Calculating total costs
4. Using technology in problem solving

Special Materials/Equipment

Calculators; reference books on pets. *Optional:* Computers and printers; Internet access; an overhead projector and transparencies for presentations.

Development

Ask your students if any of them have pets. It is likely that many of them do. Ask for volunteers to tell the class about their pets, and direct the discussion to the expenses related to them.

- Begin this project by explaining that students will work in pairs or groups of three to determine the costs of keeping a pet. They may select any pet for the project, including one like their own. (No matter which type of pet they choose, they probably have little idea of the expenses involved.)

- Distribute copies of Student Guide 54.1, and review the information with your students.

- Hand out copies of Data Sheet 54.2, "Cost Factors to Consider in Choosing a Pet." Review the sheet with your students, and point out that not all of the factors apply to all pets. Note that students should focus on the factors that are relevant to the pet they selected to research.

- Plan on reserving one or two periods in the library for research, depending on the number and kinds of references that are available on this topic. Prior to starting this project, ask your librarian to reserve books and references on pets. Suggest that students also consult online sources, which are likely to provide substantial information, and encourage them to conduct additional research on their own.

- Remind students to write a summary of their findings, including a list of all costs. Encourage them to write their reports on computers, which will make the task of revision easier. Students should also be prepared to share their findings orally with the class. You may suggest that they use an overhead projector to highlight their list of costs.

398

Wrap-Up

Students explain which pet they would select and the costs for keeping that pet. You might also wish to display their summaries.

Extension

Ask students who have pets to estimate the yearly costs of the pet and then compare these costs to the findings of the class. How close were they? What might account for any differences?

STUDENT GUIDE 54.1
The Costs of Pets

Situation/Problem

Imagine that you are thinking about buying a pet. You and your partner or group are to determine the costs of having this pet. Select any type of pet you like; then research its purchase price and all the costs necessary to keep it for a year. At the end of the project, you will summarize your findings in a brief report. You will also share your findings orally with the class.

The Costs of Pets *(Cont'd.)*

Possible Strategies

1. Consider the pet you would like to research. You may choose a common one such as a dog or a cat, or an exotic one such as a snake or tropical bird. (Remember that the less common the pet is, the harder it might be to find information.)

2. Consult reference sources on the pet you chose, and estimate the cost for keeping it for a year.

3. Consult online sources. Conducting a search for a specific pet will provide you with information. You can find Web sites that offer information about materials and equipment you will need for this pet by searching with the term "pet supplies."

Special Considerations

- For most pets, the information regarding costs will be presented in ranges or averages. Your final costs therefore will most likely be based on estimates.

- Include the purchase price of the pet, as well as all other costs. Keep in mind that some pets can be adopted for little or no cost.

- Since different pets have different needs, carefully consider the information presented on Data Sheet 54.2.

- Summarize your findings in a brief report; include an itemized list of costs and a total at the end. Write your report on a computer, which will make revision easier.

- Be prepared to explain your findings to the class.

To Be Submitted

Your report and costs.

Notes

_____

_____

_____

_____

_____

_____

Cost Factors to Consider in Choosing a Pet

Pets may be amusing, enjoyable, and satisfying; they can also be expensive. Although not all the factors below apply to all pets, many of them do. Use the ones that apply to the pet you have chosen.

- Initial cost. Some pets can cost hundreds of dollars or more.
- Any fees for a license.
- Food.
- Veterinarian fees. Many pets require routine examinations and inoculations. Include an estimate for other medical care needed during a typical year.
- Grooming fees. Some pets require regular or periodic grooming. If you do it yourself, at the least you will need bathing and grooming implements and supplies; if you take the pet to a professional groomer, you will need your wallet.
- Costs for special items—for example:

    Beds, crates, travel boxes, and litter boxes

    Food and water bowls

    Collars, leashes, chains

    Brushes, combs

    Toys, scratching posts

    Flea collars or powders

- In the case of fish you will need an aquarium, air pump, filter, carbon, heater, gravel, light, and other supplies.
- For birds you will need a cage, swing, and beak sharpener.
- For hamsters you will need a cage and running wheel.
- For pets such as lizards, salamanders, frogs, or toads, you will need a terrarium.

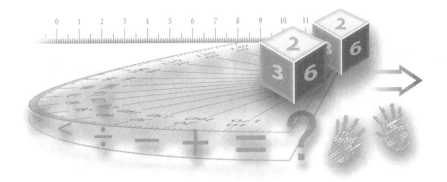

Maintaining a math class Web site

The Internet is a vast resource, a cyberstorehouse of information easily accessible with the click of a mouse. It also provides an opportunity to offer your students an ongoing math project.

Goal

Working individually or in small groups, students will maintain a math class Web site. *Suggested time:* One class period to explain the project and additional time as necessary to create material and post the material on the Web site.

Math Skills to Highlight

1. Skills will vary depending on the material students create and post to the site.
2. Using technology to communicate mathematical concepts.
3. Using technology in problem solving.

Special Materials/Equipment

A Web site; computers with Internet access.

Development

This project is wide open with opportunities. You can limit it, making overall management easier, which is a practical strategy if most of your students possess relatively weak computer skills, or you can allow students significant latitude in maintaining the site, a good choice if many of your students possess advanced computer skills. Most classes fall somewhere in between. Note that the design of this project is a general blueprint. There are many ways you can adapt the ideas here to meet the needs of your students and program.

Before beginning this project, you must establish your Web site. If your school or school district maintains a site, you may be able to secure space for your Web pages. Check with your administrator in charge of Web services or your school's computer technician. If you are unable to secure space on your school's site, consider an Internet service provider. Virtually all offer Web hosting, either including it with a subscription to the service or offering it for an additional fee. In either case, be sure to obtain the approval of your supervisor.

Setting up a site is relatively simple. Many Web site hosts offer easy instructions and templates. A template is a basic plan for the site that can be expanded as necessary. Most do not require sophisticated computer skills or programming language. Unless you have experience in designing Web sites (perhaps your own), we suggest that you start out simple with pages on which students can post brief articles about math, special problems, puzzles, and links to other interesting math sites. As you (or your students) gain experience, you can add animation, music, and interactivity. Your school's computer tech can no doubt help you set up your site. Students with savvy computer skills can also help. (Some can probably set up the entire site for you. This can be a miniproject in itself.)

Before introducing this project to your students, decide whether you will implement it with all of your students at the same time (a rather big endeavor) or with one or two classes at a time. We recommend the latter. Doing this project with several classes at once can quickly become overwhelming. Instead, if you have five classes and the school year runs ten months, you might have each class be responsible for maintaining the Web site for two months. You can, of course, adjust the classes and times in a manner that best suits your instructional needs.

You also must decide what type of material will be acceptable for the Web site. Short articles on math, problems of the day, math news, problem-solving tips, games, and puzzles are some ideas. Once you have your Web site set up and have decided how you wish to implement this project, you are ready to begin.

404

- Start by explaining to your students that they will work individually or in small groups and will, as a class, maintain a math Web site.
- Explain that it will be their responsibility to provide and update material.
- Distribute copies of Student Guide 55.1, and review the information with your students. Note the type of material they can post on the site.
- Hand out copies of Data Sheet 55.2, "Tips for Building a Web Site." Discuss the material with your students.
- Depending on the interests of your students, you might suggest that groups focus on specific materials. For example, students who like to write may prefer to contribute articles for the site. Others who like math might be inclined to create and post special types of problems. Still others who enjoy art might prefer to handle illustrations or animation. Those who are knowledgeable about computers and Web sites may assume the role of "technical experts" who help others create and post their material. In general, these students will help keep your site up and running. There will be plenty for everyone to do.
- If you intend to maintain the site from your school's computer room, reserve time for the class, and be sure to inform students when they will have access to the room. This will enable them to produce material in advance. If possible, ask your school's tech person to be present to lend a hand, at least in the early stages of the project.
- Encourage your students to create as much material outside class as possible. This will save class time.
- Be sure to check the content of any material your students produce before it is posted to the Web site.
- Make your Web site available to others. Publish the site's URL (Web address) in school and PTA publications, and send a flyer home to parents informing them of your site and asking them to visit. You might also inform the major search engines about your site. Most accept new listings; many of the listings are free. Simply go to the home page of the search engine and look for links that will enable you to register your site. To find a listing and links to search engines, go to www. allsearchengines.com.
- Explain to your students that once the Web site is up and running, material should be regularly updated. You might encourage students to do this weekly or biweekly. More than once a week may be a difficult schedule for some students; less than twice a month, and the material will become stale. Depending on your students, you may schedule class computer time and have all groups update material at the same time—perhaps every Friday—or you may encourage groups to set up their own schedules to update material.

Wrap-Up

View and discuss your class Web site periodically. Talk about how it might be improved through continued upgrading and maintenance.

Extension

Encourage students to visit other math Web sites. Explain that the ideas, design, and content they learn from other sites can help them to improve their own.

406

STUDENT GUIDE 55.1

Maintaining a Math Class Web Site

Situation/Problem

You (or you and your group) are to assume responsibility for helping to maintain your class's math Web site. You are required to regularly create and post material to the site.

Possible Strategies

1. Visit various math Web sites to learn about the kinds of material math sites contain. This background knowledge can be helpful to you as you consider the types of material you would like to create for your site.

2. Brainstorm to identify possible types of math material you might like to create.

Maintaining a Math Class Web Site *(Cont'd.)*

Special Considerations

- Select material you will enjoy working with—for example:

    Articles about math

    Problems of the day or week

    Math tips

    Problem-solving suggestions

    Puzzles and games

    Questions about math that others might be able to answer

    Reviews and links to other math Web sites.

- Avoid developing material that is overly complex and that will be difficult to post to a Web site. For example, material with complex art might be hard to create.

- Do not be intimidated by the procedures for posting material to the Web site. Most material can be posted quite easily. Refer to Data Sheet 55.2.

- Create your material outside class. Use your computer to post material to the site.

- Coordinate your material with the material of other students. Be sure to provide links to related material.

- Obtain your teacher's approval before posting anything to the Web site.

To Be Submitted

Your material on the Web site.

Notes

_____

_____

_____

_____

_____

_____

_____

Name _____

Tips for Building a Web Site

- Think of a Web site as a book, in which each link takes you to another page of a book. Instead of turning a paper page, though, viewers turn pages by clicking a mouse.

- Select manageable material for your site. Avoid trying to create and post complex material until you have gained some experience in working with Web sites.

- Decide how many pages you will need for your material. If you plan to create and post a problem of the day, you will probably need only one page or a partial page. If you are writing an article with tables and charts, you will need more.

- If your material requires several pages, first map it as a flowchart. Use index cards as the parts of the chart, arranging them to give you an idea of how to set up your material. Different index cards can represent different parts of your material that will be connected by links.

- If your Web site host offers instructions of how to build a site and post material on it, be sure to follow the guidelines. Step-by-step instructions can make posting material relatively simple.

- Create as much of your material in advance as you can. Save the material as a separate file on a disk or CD; then insert it to the site. Trying to create material while you are working on the site can be difficult.

- Consider highlighting your material with clip art. Copyright-free clip art is included with many word processing programs. It can also be obtained from clip art CDs and Internet sites. In many cases, you can insert art directly to the Web site. Always be sure that any clip art you use is free.

- If you are unsure how to post material on your site, ask for help. Your teacher, your school's computer technician, or another student skilled with computers will likely be able to help. In many cases, what may at first seem to be an impossible problem is in fact minor and is easily corrected.

- After you post your material on your site, be sure to update it regularly.

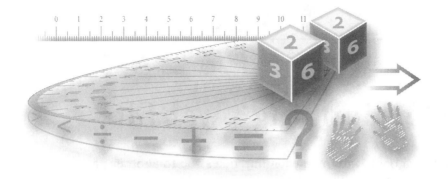

Selecting a sound system using the Internet

Not too long ago, most people characterized the Internet as a storehouse of information on just about every topic imaginable. While the Internet remains a storehouse of information, it has also evolved into a storehouse of products and merchandise. You can buy just about anything from cyberstores, often for lower prices than what you can find locally. This project provides students with the opportunity to use the Internet in a quest to purchase a sound system for the school's music room.

Goal

Working in pairs or groups of three, students will research sound systems on the Internet, determining which would be the best system to buy. They will assume they have $1,500 to spend. Students will present their results to the class. *Suggested time:* Two class periods—a partial period to introduce the project and a period or so to present results. Some of the research will probably need to be done outside class.

Math Skills to Highlight

1. Comparing and tallying costs
2. Making decisions based on cost
3. Using technology in problem solving

Special Materials/Equipment

Computers with Internet access. *Optional:* An overhead projector and transparencies for presentations.

Development

Ask your students how many of them have sound systems (also commonly referred to as stereo or audio systems) in their rooms or homes. It is likely that most do. (It is also likely that many of your students know far more about these systems than you do.) Ask them to name some of the features they look for in a quality system.

- Begin this project by explaining to your students that they are to imagine that they have $1,500 to spend on a new sound system for the school's music room and that they are to research the system on the Internet. They must determine the features they would like in a system, then find a system that satisfies their requirements but stays within their price range.

- Distribute copies of Student Guide 56.1, and review the information with your students. Note that students are free to research a variety of Web sites, but remind them to record each site's URL in case they need to return to the site for more information.

- Explain that students are to do much of the research on their own. They may use their own computers or reserve time in the school's computer room. You might want to reserve time for the class in the computer room to get your students started.

- Remind students that although they are free to select the system and components they wish, they must choose equipment that is compatible and makes a complete system.

- Hand out copies of Worksheet 56.2, "The Best Sound System," and review the material with your students. Note that they are to complete the worksheet as a part of the project.

- Remind students to be prepared to discuss their selection on completion of the project. You may suggest that they use an overhead projector to show their results.

Wrap-Up

Students present their results to the class. It is likely that the types of systems they select will vary widely.

Extension

Suggest that students research other products on the Internet; for example, cell phones, computer systems, or wide-screen or plasma TVs.

STUDENT GUIDE 56.1

Selecting a Sound System Using the Internet

Situation/Problem

Imagine that you and your partner or group have $1,500 to spend for a new sound system for the school's music room. You are to research various sound systems on the Internet and select the best one that remains within your budget. You are to complete Worksheet 56.2 and present your findings to the class.

Possible Strategies

1. Brainstorm the features you would like in a sound system. List these features, which will be your criteria for comparing systems.
2. Visit several Web sites, and compare equipment and prices.

Special Considerations

- To find helpful Web sites, conduct a search using terms such as "sound systems" or "audio systems." Be aware of useful links to other sites.
- Do not spend time browsing on Web sites that do not offer the information you need. Focus on sites that do.
- Consider systems that satisfy your criteria of a quality sound system, but be ready to compromise, if necessary, because of price.
- Be aware that you can sometimes purchase a basic system and then add other components. Sometimes this can save money without sacrificing quality.
- After selecting a system, complete Worksheet 56.2. Be prepared to present your results to the class.

To Be Submitted

Your completed worksheet.

Name _____

The Best Sound System

Complete this form on the sound system you selected.

Web Merchant and URL (Web address): _____

Name of Sound System: _____ Model Number: _____

Components of System: _____

_____

_____

_____

Features of System: _____

_____

_____

_____

Why You Selected This System over Others: _____

_____

_____

_____

Total Cost: $ _____

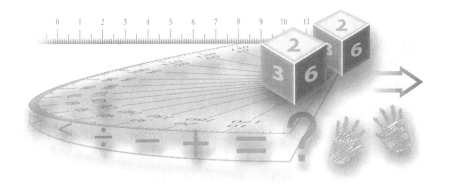

Buying a car

As early as middle school, many students look forward to the day they will be able to drive. Few, however, think about the responsibilities and expenses that come with driving. This project enables students to see the costs related to buying and owning a car.

Goal

Working individually, students will assume that they are shopping to buy a car. They will read advertisements about cars and select a model that they feel best suits their needs and budget. Students will calculate the expense of buying and owning the car, including financing, operating, and maintenance costs. At the end of the project, they will explain their selection and their anticipated costs during a class discussion. *Suggested time:* Two class periods, although students will likely spend time outside class researching information about cars.

Math Skills to Highlight

1. Anticipating and estimating costs
2. Tallying various cost factors
3. Understanding the terms *financing, principal, annual percentage rate (APR), finance charge, down payment,* and *monthly payments*
4. Using a formula to find the finance charge on a loan
5. Making decisions based on cost
6. Using technology in problem solving

Special Materials/Equipment

Calculators; the classified and automotive sections of local newspapers. *Optional:* Articles comparing the values of specific cars in such publications as *Consumer Reports* and *Car and Driver;* computers with Internet access for research; an overhead projector and transparencies for presentations.

Development

Ask your students how many of them expect to own a car someday. If you teach high school, perhaps some of your students already do. Of those who do not, most will enthusiastically raise their hands. Now ask if they have ever thought about the overall costs associated with owning and operating a car. Most probably have not. Aside from the obvious costs such as purchase price, gas, and maybe insurance, most students fail to anticipate the actual costs of car ownership.

- Begin the project by explaining that students will work individually and imagine that they are ready to buy a car. Knowing their financial status (or anticipating what their finances are likely to be), they are to select a car that fits their budget and their needs.

- Explain that students should look through the automotive and classified sections of local newspapers to find a car they realistically can afford. Ask students a few days before the project to bring in those parts of newspapers from home. If you also start collecting materials ahead of time, you should have plenty for the project.

- Publications such as *Consumer Reports* and *Car and Driver* often contain articles about cars. Students may wish to consult these and similar resources to find information about the cars they are considering. If such publications are not available in your school library, your local library may have them. You might also suggest that students consult online sources where they will find substantial information on cars.

- Distribute copies of Student Guide 57.1, and review the information with your students. Especially review the steps necessary for finding the approximate finance charge and monthly payments for car loans. You may need to provide your students with some examples.
- Discuss the terms *financing, principal, annual percentage rate, finance charge,* and *monthly payments.*
- Hand out copies of Data Sheet 57.2, "Car Costs." Review the material with your students, and point out that the costs listed on the sheet are estimates and will vary around the country. Students may use the costs on the data sheet for this project. If students own cars and are already familiar with the costs, they should use the actual expenses.
- Hand out copies of Worksheet 57.3, "A Car Buyer's Cost Sheet." Explain that students are to list and tally all of their expenses on the worksheet.
- Remind students to be prepared to discuss their selection and anticipated costs on completion of the project. You may suggest that they use an overhead projector in their presentations. They can make a transparency of Worksheet 57.3 to share their results.

Wrap-Up

Conduct a class discussion in which students share the information they compiled on their worksheets.

Extension

Suggest that students compare leasing a car to buying one. What are the advantages of each method? What are the disadvantages? Provide class time for students to discuss their findings.

STUDENT GUIDE 57.1

Buying a Car

Situation/Problem

You are to imagine that you are shopping for a car. You are trying to buy a car that meets your needs but also satisfies your financial status. In selecting a car, you are to calculate your overall anticipated expenses, including the purchase price, finance charge (if any), and operating and maintenance costs. On completion of this project, be prepared to explain your choice to the class.

Possible Strategies

1. Consider the type of car you would like.
2. Consider your financial condition. Be realistic. Choose a car that you think you can afford.
3. Consult automotive and classified advertisements in trying to find a car that will satisfy your needs and remain within your budget. You may also consult online sources for information.

Buying a Car *(Cont'd.)*

Special Considerations

- Are you interested in a new or used car? New cars cost more.

- Use search terms such as "new cars" and "used cars" to find sites about cars.

- Will you pay cash for your car, or will you need to obtain a loan? Financing your car (taking a loan) allows you to put down less cash, but you will need to make monthly payments that include interest. Interest will increase the cost you pay for the car. You can find the approximate finance charge you would have for a loan by using the following formula:

$$\text{Finance Charge} = \frac{A(N + 1)(APR)}{2P}$$

  A = amount of money borrowed (principal)
  N = total number of payments
  APR = annual percentage rate (interest; for this project use an APR of 7%)
  P = number of payments per year

- After you find the approximate finance charge, add it to the amount of your loan. This is how much money you will need to pay back. Now divide this total by the number of payments. The answer equals the amount of your monthly payments. To find the amount of your total payments for each year of the loan, simply multiply your monthly payments by 12. (A loan for one year will have 12 payments. A two-year loan has 24, a three-year loan has 36, and a four-year loan has 48. The longer you take a loan, the lower your monthly payments will be. However, your finance charge will be greater.)

- Carefully review the information and costs on Data Sheet 57.2. Use this information for calculating your operating and maintenance expenses.

- Use Worksheet 57.3 to list and total your costs. Remember to find the yearly cost of operating and maintaining your car.

- Be prepared to discuss your choice of car and anticipated costs with your class after the project is finished.

To Be Submitted

Your worksheet.

DATA SHEET 57.2

Car Costs

Following are common expenses in operating and maintaining a car. Include the ones that apply to your car on Worksheet 57.3. Note that the costs are estimates and will vary for different parts of the country.

- Payment for any loan, including finance charges.
- Insurance (for a new driver):

  Small car, $2,000 per year

  Midsized car, $2,300 per year

  Sports car, $3,000 per year

  SUV, $3,400 per year
- Fees for license and registration, $75.
- Gasoline. Use $2.99 per gallon for estimating your fuel costs. To find your estimated cost for gas for the year, do the following:

  Estimate the total number of miles you expect to drive for the year. You can estimate each week's total miles and multiply by 52.

  Divide the total number of miles by the car's rated mpg (miles per gallon). If you do not know the mpg of your car, use these estimates: small car, 40 mpg; midsized car, 25 mpg; sports car, 22 mpg; SUV, 20 mpg.

  Multiply this answer by the cost of gasoline.
- Costs for tune-ups, oil changes, and maintenance:

  Small car, $250 per year

  Midsized car, $450 per year

  Sports car, $650 per year

  SUV, $700 per year
- Include any additional costs such as CD player, enhanced interior, sun roof, or special tires or wheels.

Name _____

A Car Buyer's Cost Sheet

List all of the expenses you expect in purchasing, operating, and maintaining the car of your choice. Operating and maintenance costs should be totaled for the year.

Model and Year of Car _____

Purchase Price $_____

OPERATING AND MAINTENANCE COSTS

Description of Expense	Cost
Total Yearly Cost	

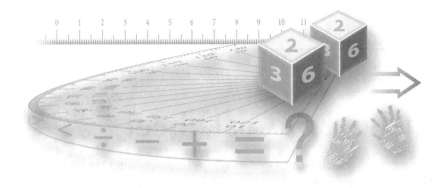

What is on the test?

Teachers always give students tests. For a twist, this project gives students the chance to give each other tests that they themselves write. Allowing students to write a test that they feel underscores the concepts and skills in a unit is an excellent reinforcement activity.

Goal

Working in groups of four or five, students will write a test on the current unit of study. They must also provide an answer key and point distribution. *Suggested time:* Two class periods.

Math Skills to Highlight

1. Specific skills will vary depending on the unit of study and the skills students select.
2. Using technology in problem solving.

Special Materials/Equipment

Will vary depending on unit of study. *Optional:* Computers and printers for use in writing the tests; software such as Microsoft Equations Editor and Design Science's Math Type, which will enable students to include math symbols and equations on their tests.

Development

As you near the end of a unit and announce the upcoming test, ask your students to identify the important concepts and skills they have learned. They should list them individually. Let students keep their lists, and suggest that they add any other ideas as they think of them. They will refer to their lists for the project.

- Begin the project by explaining that students will work in groups to write a test for the current unit of study. They will also be required to provide a key. On completion of the project, you will review the tests and select two that you will administer to the class. You will need two tests so that the groups whose tests you select do not receive their own. Be sure not to divulge whose tests you are using so that students are not tempted to share answers in advance. Emphasize that you reserve the right to add or delete material; however, assure students that the two tests you judge most appropriate will be given to the class.

- Distribute copies of Student Guide 58.1, and review the information with your students. Emphasize that each group should divide the tasks and that all members should contribute to the test.

- Point out that the tests must assess all of the concepts and skills covered in the unit, preferably by a variety of questions.

- Note that the groups should include a point distribution for their tests. You may need to explain how students may decide on the value of each question or problem.

- Hand out copies of Data Sheet 58.2, "Elements of a Good Math Test," and review the material with your students. The data sheet offers information that will help students design effective tests.

- Encourage students to use computers when writing their tests. If their software does not support writing formulas or numbers (for example, fractions), suggest that they leave a space and later write those problems in with black pen or a heavy pencil.

- Emphasize the importance of accuracy. Answer keys must be correct.

- Set a deadline for completion of the groups' tests at least two days prior to the testing date. That will give you time to review the tests, make any adjustments (if any are needed in your opinion), and make copies for the class.

- After you collect the tests, be sure to review them carefully and check the answer keys, especially for the two you are planning to use.

Wrap-Up

Students will take the tests created by their classmates.

Extensions

On the day after the test, ask your students to write their thoughts and comments about the tests they took. Collect the papers and share them with the groups of students who wrote the tests.

STUDENT GUIDE 58.1

What Is on the Test?

Situation/Problem

Your group will write a test for the unit your class is currently studying. You will focus the questions on the key concepts and skills presented in the unit and include an answer key and point value for each problem. On completion of the project, your teacher will select what he or she feels are the two best tests and administer them to the class. (Two tests will be given so that the groups who designed the tests will not take their own.) Your teacher reserves the right to add or delete material if he or she feels it is necessary.

Possible Strategies

1. Each member of the group should write a list of what he or she feels are the important concepts and skills covered in the unit. Compare the lists and identify those that all of you feel are the most important.

2. Discuss how to organize the test. What types of problems should it have? How many parts? How long should it be?

3. Divide the tasks of creating the test among group members. Perhaps each member may assume responsibility for a portion of the test.

What Is on the Test? *(Cont'd.)*

Special Considerations

- Think of the types of problems you have worked on in your study of this unit. The test should reflect those problems and the concepts and skills they contain.
- Think about the math tests you have had in the past. Use them as a guide for designing this test.
- Consult Data Sheet 58.2 for suggestions on creating a test.
- Be sure that any directions you provide on your test are clear.
- Leave enough space on the test for students to work out problems.
- Estimate the time students will need to complete the test. Your test should not be so long that students will have trouble finishing it during class. Neither should it be so short that they will complete it in a few minutes.
- Avoid trying to create extremely difficult problems just to stump other students. Also avoid making the test too easy. Try sample problems on some of the members of your group.
- Write your test on a computer. If your software does not allow you to write math symbols or number problems, leave a space on the page and write the problem or symbol in later with a dark pen.
- Remember to include an answer key and a point distribution.

To Be Submitted

A copy of your test and answer key.

Notes

_____

_____

_____

_____

_____

_____

_____

Name _____

Elements of a Good Math Test

The best tests are those that focus on the important skills and concepts of a unit. Following are suggestions of how to create an effective math test.

- A test should assess what has been taught.
- A test should have different kinds of questions, including:

    True or false

    Multiple choice

    Short answer, such as a number, estimate, or term

    Finding solutions to various kinds of problems

    Open-ended questions or problems that students must explain, or write a brief description of how they would solve the problem

- A test should provide students with problems that require them to choose one or more of several strategies for finding a solution, including:

    Guessing and checking

    Making a table

    Looking for a pattern

    Making a list

    Making a simpler problem

    Working backward

    Drawing a diagram

    Writing an equation

- A test should allow for the use of calculators and such items as rulers, protractors, and compasses, depending on the unit of study.
- A test may contain an optional bonus or extra-credit problem that relates to the content but is challenging to most students.

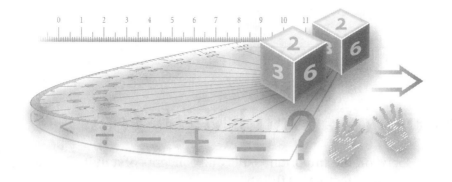

Checks and balances

The importance of keeping a balanced checking account cannot be understated. Overdrawn accounts may result in a bad credit rating. For this project, students will have the chance to write checks and maintain a running balance.

Goal

Working individually, students will be provided with practice checks, a check register, and deposit forms. They will be given a list of expenses and deposits and will be required to maintain a running balance. *Suggested time:* Two class periods.

Math Skills to Highlight

1. Tallying costs
2. Maintaining a checkbook balance
3. Understanding debits and credits
4. Using technology in problem solving

Special Materials/Equipment

Calculators.

Development

Discuss the purpose of checking accounts with your class, and explain how they work. You may also wish to note the consequences of not maintaining an accurate checking account.

- Begin the project by telling your students that they will receive materials for a practice checking account and will work individually to maintain a balance for their account. Their starting balance will be $500. Emphasize the importance of not spending more money than their account contains; if they do, their check will bounce.

- Distribute copies of Student Guide 59.1, and review the information with your students. Be certain that students understand what they are to do.

- Hand out copies of Data Sheet 59.2, "Keeping a Balanced Checking Account." Note that the sheet offers specific steps for writing checks, completing deposit forms, and maintaining a check register. Depending on your class, you may wish to offer some examples before letting students begin the project.

- Hand out copies of Data Sheet 59.3, "Debits, Credits, and ATMs." This sheet contains various expenses and deposits that students are to use in maintaining their checking accounts. Emphasize that of the various debits, students are to pick five for which they will write checks. They must record the ATM withdrawal. They will also need to fill out two deposit forms.

- Distribute two copies of Worksheet 59.4, "Practice Checks," to each student. Since each sheet contains three checks, each student will have six checks. Instruct your students to number the checks consecutively from 101 to 106 and write their names at the upper-left-hand corner of each check. They will use the checks to pay for the expenses contained on Data Sheet 59.3.

- Distribute copies of Worksheet 59.5, "A Check Register and Deposit Forms." Explain that the register is the place where an individual keeps a record of the checks written and debits and deposits made to the account.

- If you would like to expand the project, provide students with more copies of Worksheets 59.4 and 59.5. Students may obtain additional "items to buy" from local newspaper advertisements.

Wrap-Up

Have students exchange their worksheets with a partner who, using a calculator, checks their account, making sure that it is balanced correctly. Collect the worksheets, and discuss students' impressions of maintaining a checking account. What did they find hard? What did they find relatively easy?

Extension

Invite a representative of a bank to class to discuss checking accounts, as well as some of the other types of accounts banks offer.

Checks and Balances

Situation/Problem

You are to maintain a practice checking account. You will write checks, complete deposit forms, and keep a check register. On completion of the project, you will exchange your materials with a partner and check each other's math. Be prepared to share your thoughts about checking accounts with the class during a discussion.

Possible Strategies

1. Review Data Sheet 59.3, and pick six items you will write checks for. Note that your account has a prior balance of $500.

2. As you write your checks, subtract the amount of each from your balance in your check register (Worksheet 59.5). This will reduce the chances that you will overdraw your account.

3. As you complete deposit forms, remember to add the sum to your register.

Checks and Balances *(Cont'd.)*

Special Considerations

- Consult Data Sheet 59.2 to learn how to write checks, fill out deposit forms, and maintain a check register. Be accurate with your math.
- When you receive your practice checks (Worksheet 59.4) and register and deposit forms (Worksheet 59.5), write your name at the upper-left-hand corner. Number your checks 101 through 106.
- Remember that checks are subtracted from your balance. These are called *debits*. ATM (automatic teller machine) withdrawals are also deducted from your balance. Deposits are added and are called *credits*.

To Be Submitted

Your completed worksheets.

DATA SHEET 59.2

Keeping a Balanced Checking Account

How to Write a Check

- Use a pen, and write clearly. Do not cross out or change anything on the check once you have written it.

- Write the date (the day the check is written).

- Write the payee's name (the person to whom the check is written) after "Pay to the order of."

- Write the value of the check next to the dollar sign. Use a decimal point for writing change; for example, $29.99.

- On the line below the payee's name, write the amount of the check in words. Start at the left, at the beginning of the line. Begin with a capital letter and write out the dollar value; then show any change by using the word *and* and the value over 100, for example, "twenty-nine and $\frac{99}{100}$." If any space to the right remains, draw a line from the words or numbers to the word *dollars* at the extreme right.

- The memo is optional. You may write what the check is for here.

- Sign the check.

How to Complete a Check Register

- Write the check number.

- Write the date of the check.

- Write to whom the check was written.

- Write the purpose of the check.

- Write the amount of the check. Subtract the amount of the check from the balance in your account.

- Record any ATM withdrawals, and subtract them from the balance. Remember to include any ATM fees.

- When recording a deposit, write the date and the amount of the deposit, and add it to the balance in your account.

How to Fill Out a Deposit Form

- Write the date.

- Put the amount of money in bills, coins, and/or checks.

- Write the total.

Debits, Credits, and ATMs

Select six debits from the following list, and write checks for them. The company, organization, or person to whom you are to write the check is in parentheses. The list is arranged by date, and you must choose the items in order. You must also record the ATM withdrawal. Note that some deposits are included for which you must fill out deposit slips. Be sure to maintain a balanced checking account. You have a prior balance in your account of $500.00.

April 3	You buy a new camera, $239.95. (Flash Camera Shop)
April 6	You buy a new pair of sneakers, $109.95 (High Five Sports)
April 10	You receive $20.00 for helping your neighbor clean out his garage. You make a deposit.
April 17	You pay $48.00 to have your bike repaired. (Tom's Bike Shop)
April 21	You buy new in-line skates, $99.99. (The Skate Place)
April 28	You decide to put the money you have been hoarding from your paper route into your checking account. You make a deposit of $93.00.
May 2	You buy your best friend a birthday gift, $29.89. (Highway Gift Shop)
May 8	You pay your mother the $25.00 you borrowed. (Your Mother)
May 16	Out of cash, you head for the nearest ATM and withdraw $60.00. (The $2.00 fee for using this ATM is deducted from your account.)
May 20	You purchase your favorite group's new CD for $19.95. (Great Sound Music Store)
May 28	Your money for the upcoming class trip is due. You must pay $18.00. (Your School)

Practice Checks

_____ 20 _____

PAY TO THE
ORDER OF _____ $ _____

_____ DOLLARS

South River Office
South River, NJ 08882

FOR _____ _____

_____ 20 _____

PAY TO THE
ORDER OF _____ $ _____

_____ DOLLARS

South River Office
South River, NJ 08882

FOR _____ _____

_____ 20 _____

PAY TO THE
ORDER OF _____ $ _____

_____ DOLLARS

South River Office
South River, NJ 08882

FOR _____ _____

Name _____

A Check Register and Deposit Forms

						Balance
Check Number	Date	Description of Transaction	Payment Debit	✔	Deposit/ Credit	

DATE _____ 20 _____

	DOLLARS	CENTS
CURRENCY		
COIN		
CHECKS List Each Separately 1		
2		
3		
4		
5		
6		
7		
8		
9		
10		
Total		

DATE _____ 20 _____

	DOLLARS	CENTS
CURRENCY		
COIN		
CHECKS List Each Separately 1		
2		
3		
4		
5		
6		
7		
8		
9		
10		
Total		

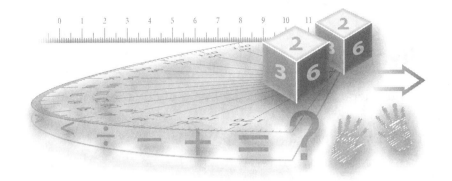

Math in my life: An assessment

At the end of the year, math students who have worked on a variety of projects to supplement their regular curriculums frequently grow in their understanding of and attitudes toward mathematics. They come to see math as not simply a subject taught in school, but as a significant factor in their lives. Because this project assumes that students have had experience working on math projects, it is best assigned near the end of the year.

Goal

Working individually, students will reflect on their attitudes and feelings about math, in particular on the class of which they were a part this year. Students will have the opportunity to create a symbol, design, illustration, or model that represents how their ideas about math have changed or grown. *Suggested time:* Two class periods over the course of a week.

Math Skills to Highlight

1. Skills will vary depending on the students' individual responses to the project.

2. Using technology in problem solving.

Special Materials/Equipment

Will vary depending on the students' needs; however, rulers, scissors, felt-tipped pens, markers, colored pencils, graph paper, poster paper, and similar products should be made available. *Optional:* Computers and printers.

Development

Reflect with your students on some of the projects or mathematical highlights you have shared during the year. You might also briefly review some of the topics you have studied. This will help to focus students' thoughts on the past year.

- Begin the project by telling students that they will work individually to assess their attitudes and feelings about math, and especially how they might have changed during this year. Making periodic assessments of one's progress in a specific subject is a good way to keep track of growth. It helps to uncover how much progress has been made and enables a person to adjust his or her goals.

- Distribute copies of Student Guide 60.1, and review the information with your students. Note that the student guide offers questions and suggestions that may help students in their self-assessment.

- Encourage students to think of a symbol, design, illustration, model, or some other way—perhaps a poem or song—that will represent how they feel about mathematics, and especially how they have grown in their appreciation and understanding of math. While they may use the materials you provide in class, suggest that they also use materials they have at home.

- Give students time in class to begin. You may wish to let them brainstorm with friends, but insist that they work alone on their assessments. Depending on their representations, some students may also work at home.

Wrap-Up

Students show and explain their creations to the class. Display their work.

Extension

Conduct a class discussion on the ways students can get the most out of a math class. You will find that most have some good ideas and helpful suggestions.

STUDENT GUIDE 60.1

Math in My Life: An Assessment

Situation/Problem

You have worked on various topics and math projects this year. Certainly you have grown in your skills and understanding of math. Perhaps your perceptions, impressions, and attitudes about math have also changed. For this project, you are to examine your feelings about math and create a symbol, design, illustration, or model that represents them.

Possible Strategies

1. Think about this past year in math. On a sheet of paper, list some of the highlights and disappointments.

2. Consider if your attitudes about math have changed. Write down your feelings.

3. Think about a way you can represent your feelings.

Math in My Life: An Assessment *(Cont'd.)*

Special Considerations

- To get in touch with your thoughts and feelings, ask yourself questions like the following:

  How relevant is math in my life?

  What new skills did I learn in math this year?

  What gave me the most trouble?

  What would I like to learn more about in math?

  What kinds of real-life situations can math help me in?

  What project did I enjoy the most this year?

  What math skills are my strongest?

  What will I probably remember most about this year in math?

  How important would I rate math in my life? Very? Average? A little?

- Be creative in selecting a method to represent your feelings. You may use a symbol of some sort, an illustration, a model, drawing, or something else. Perhaps you might like to write a song. Whatever you choose, it should show the importance of math in your life.

- Consider designing your symbol using a computer.

- At the end of the project, be ready to show your work and explain your feelings.

To Be Submitted

Your work.